THE

LIFE SCIENCE

BOOK

*A Comprehensive Warm-Up
for Grades 8-12*

Jude Nnanabu, Ph.D.

ISBN-10: 1478177063
EAN-13: 9781478177067

THE
LIFE SCIENCE
BOOK

A Comprehensive Warm-Up
for Grades 8-12

Jude Nnanabu, Ph.D.

Dr. Jude Nnanabu has spent many years teaching a subject he truly enjoys: science. An innovator who believes in results, he has worked to improve schools and has researched ways to guide students to academic success.

Thanks to my wonderful wife, Letitia Reid-Nnanabu, who helped to organize, proofread, and type this book.

TABLE OF CONTENTS

This science warm-up will help the teacher get students started on the thinking process, just like warming up a car before driving it away. Warming up can also help reinforce any science teaching, and help prepare the student for learning.

LIFE SCIENCE

1. What is science?

2. What is life science?

3. How do science and technology influence our lives?

4. What is a scientist?

5. What do scientists use to predict what will happen?

6. How are scientists like detectives?

7. What are the skills used in science?

8. What is a science journal?

DOING SCIENCE EXPERIMENTS & SOLVING SCIENCE PROBLEMS

Steps of the scientific method

- Ask a question
- Do background research
- Construct a hypothesis
- Test your hypothesis
- Analyze your data and draw a conclusion
- Communicate your results

Overview of the scientific method

The scientific method is a process of experimentation that we use to explore observations and answer questions. Scientists use the scientific method to search for cause and effect relationships in nature. In other words, they design experiments so that changes to one thing cause something else to vary in a predictable way.

The scientific method teaches students to construct a hypothesis, design, execute, and evaluate their experiments just like professional scientists do.

STEPS OF THE SCIENTIFIC METHOD

Ask a question

The scientific method begins with a question about something that you observe: how, what, when, who, which, why, or where? The scientific method is best at answering questions about something that you can measure, preferably with a number or something else. For example, "What would happen to an ice cube if I placed it in a saucer of water?"

Do background research

Rather than starting from scratch to try to answer your question, you want to be a savvy scientist. Use library and Internet research to help you find the best way to do things and ensure that you do not repeat mistakes that others have already made.

Construct a hypothesis

A hypothesis is an educated guess about how things work. You are required to state your hypothesis in terms of something that you can easily measure and that should help you answer your original question. Try using the "If_____ then_____" construction. For example, a hypothesis could be "If I place an ice cube in this saucer of fresh water, then it will melt in one hour."

Test your hypothesis by doing an experiment

The testing you do in your experiment may either disprove your hypothesis or appear to confirm it. It is important for your experiment to be a fair test. You conduct a fair test by making sure that you change only one factor at a time while keeping all other conditions the same. You should also repeat your experiments several times to make sure that the first results were not just an accident.

Analyze your data and draw a conclusion

Once your experiment is complete, you collect your measurements and analyze them to see if your hypothesis is true or false. Scientists often find that their hypotheses are false, so then they construct a new hypothesis, starting the entire process of the scientific method over again. Even if they find that a hypothesis appears confirmed, they may want to test it again in a new way.

Communicate your results

Your science experiment or science fair project is not complete until you communicate your results to others in a final report or other public format (such as on a display board). Professional scientists do almost exactly the same thing by publishing their final reports in scientific journals or by presenting results on a poster at a scientific meeting.

Guiding questions for understanding the scientific method

1. What is a variable?

2. What is an independent variable?

3. What is a dependent variable?

4. What is a problem?

5. What is an observation in an experiment?

6. What is a conclusion in an experiment?

7. What is an interactive method?

8. What is the scientific method?

9. Name the four steps scientists use when designing an experiment to solve a problem.

10. What is a microscope?

11. Draw and label the parts of a microscope.

12. What does a compound light microscope do?

13. What is a stereo microscope used for?

14. What can an electron microscope do?

15. What do transmission electron microscopes do?

16. What is the function of a scanning electron microscope (SEM)?

17. Who invented the microscope and when?

18. Explain why the invention of the microscope was important in the study of cells.

19. How has the development of different microscopes helped scientists to study cells?

QUESTIONS ABOUT THE SCIENTIFIC METHOD

1. Which of the following is not a process that scientists use to generate hypotheses?

   a) using informed creative imagination
   b) using prior knowledge
   c) using logical inference
   d) using their feelings about the surroundings

2. Information gathered from observing a tree that grows 2 cm over a two-week period   results in:

   a) variables   c) a hypothesis
   b) inferences   d) data

3. A controlled experiment allows scientists to isolate and test _____.

   a) several variables   c) a mass of information
   b) a conclusion   d) a single variable

4. Scientific hypotheses must be proposed in ways that _____.

   a) enable them to be proved valid
   b) ensure that an experiment will be valid
   c) enable them to be tested
   d) do not contradict previous hypotheses

5. Any _____ must have the ability to be reproduced.

a) theory   c) law
b) hypothesis   d) experiment

6. Scientists publish the details of important experiments so that _____.

a) their experimental work procedures can be reviewed
b) their work can be repeated
c) others can try to reproduce the results
d) all of the above

QUESTIONS ABOUT MICROSCOPES

1. To observe a small living organism, a scientist might use a(n) _____.

   a) TEM
   b) electronic balance
   c) compound light microscope
   d) electron microscope

2. An instrument used to separate cell parts according to density is the _____.

   a) electron microscope
   b) compound light microscope
   c) blender
   d) centrifuge

3. An instrument that allows light to pass through a specimen and that uses two lenses to form an image is a(n) _____.

   a) electron microscope
   b) compound light microscope
   c) SEM
   d) TEM

4. If the eyepiece lens of a microscope has a power of 10x and the objective lens has a power of 43x, then the _____ _____ is 430x.

5. Instead of a standard lens, the electron microscope uses a _____ to bend electrons.

6. The surfaces of whole objects can be examined with a _____ electron microscope.

7. The scientist who invented the first compound micro-scope was _____.

8. The scientist who concluded that all animals are made up of cells was _____.

9. The scientist who called box-like structures in cork "cells" was _____.

10. The scientist who observed that every cell comes from a cell that existed before was _____.

SCIENCE & SAFETY PROCEDURES

1. Which of the following is NOT regarded as a safety procedure?

   a) follow your teacher's instructions
   b) follow the textbook directions exactly
   c) read all the steps in your activity before doing it
   d) if in doubt about any part of an activity, trust your instincts

2. You may come in contact with organisms you cannot see. What safety process must be followed?

   a) open the windows of the laboratory
   b) read over the activity procedures
   c) wash hands thoroughly after completing the activity
   d) avoid wearing long sleeves

3. Safety procedures are important when working _____.

   a) with animals and organisms
   b) in a laboratory
   c) in the field
   d) all of the above

4. List ten safety procedures used in the classroom.

THE SCIENCE OF BIOLOGY

1. Biology is the study of _____.

   a) living things
   b) the land, water, and air on Earth
   c) animals and plants only
   d) the environment

2. The work of scientists begins with _____.

   a) careful observations   c) creating experiments
   b) testing a hypothesis   d) drawing conclusions

3. Which of the following terms includes all the other names?

   a) botanist   c) zoologist
   b) biologist   d) ethologist

4. Which of the following is not the outcome of science?

   a) explanation of events in the natural world
   b) investigation and understanding of the natural world
   c) the use of data to support a particular point of view
   d) the use of derived explanations to make useful predictions

5. The basic unit of mass in the International System of Units (SI) is the _____.

a) ounce   c) liter

b) meter   d) gram

6. In the metric system, the basic unit of length is the

_____.

a) kilometer   c) millimeter

b) centimeter   d) meter

7. The unit of measurement not in the International System is ____.

a) ounce   c) liter

b) meter   d) gram

8. What is the metric system?

9. Explain the difference between inference and observations.

10. Explain how SI benefits scientists in different parts of the world.

11. The SI unit used to measure liquids is the _____.

a) liter   c) gram

b) meter   d) degree

THE CHEMISTRY OF LIFE

1. An example of an organic compound is _____.

    a) NO_2 c) H_2O
    b) $C_6H_{12}O_6$ d) O_2

2. The organic molecule that releases the largest amount of energy is _____.

    a) water   c) lipid
    b) carbohydrate   d) inorganic

3. Water is a(n) _____ molecule.

    a) organic   c) carbohydrate
    b) lipid   d) inorganic

4. The nucleus is made up of _____.

    a) electrons and neutrons
    b) protons and electrons
    c) protons and neutrons
    d) protons, neutrons, and electrons

5. Three particles that make up an atom are _____.

    a) neutrons, isotopes, and electrons
    b) protons, neutrons, and isotopes
    c) positives, negatives, and electrons
    d) protons, neutrons, and electrons

6. A covalent bond is formed as the result of _____.

    a) sharing electrons
    b) transferring electrons
    c) transferring protons
    d) sharing protons

7. Suspensions are mixtures _____.

    a) in which the components are evenly distributed
       throughout the solution
    b) of water and undissolved material
    c) both a and b
    d) neither a nor b

8. A solution is a(n) _____.

    a) chemical reaction
    b) breaking of a chemical bond
    c) evenly distributed mixture of two or more substances
    d) combination of two or more liquids

9. Which of the following organic compounds is the main source of energy for living organisms?

   a) lipids
   b) carbohydrates
   c) nucleic acids
   d) proteins

10. Which one of the following is not a function of proteins?

   a) to help fight disease
   b) to store and transmit heredity
   c) to control the rate of reactions and regulate cell processes
   d) to be used to form bones and muscles

11. The process that changes one set of chemicals into another chemical is called _____.

   a) adhesion
   b) cohesion
   c) chemical reaction
   d) dissolving

12. Chemical reactions that release energy _____.

   a) do not happen
   b) do not explode
   c) always explode
   d) often happen spontaneously

13. A substance that speeds up a chemical reaction is called a(n) _____.

    a) lipid
    b) catalyst
    c) molecule
    d) element

14. The most abundant compound in most living organisms is _____.

    a) water
    b) carbon dioxide
    c) sodium chloride
    d) sugar

15. When sodium is dissolved in water, the water is called a _____.

    a) solution
    b) reactant
    c) solute
    d) solvent

16. A substance with a pH of 6 is called a(n) _____.

    a) base
    b) acid
    c) acid and a base
    d) neither an acid nor a base

17. The space surrounding the nucleus of an atom contains
    _____.

   a) electrons
   b) protons
   c) neutrons
   d) icons

18. The electrons available to form bonds are called an
    atom's _____.

   a) nucleus
   b) valence
   c) ionic
   d) covalent

19. What type of ion forms when an atom loses elec-
    trons?

   a) negative
   b) neutral
   c) positive
   d) neither positive or negative

20. If an atom contains fifteen protons, it must contain fif-
    teen _____.

21. The stomach produces a(n) _____ to digest food.

22. The pH scale is a measurement system that indicates the concentration of ____ in a solution.

23. List two essential roles that enzymes play in cells.

24. What is a molecule?

25. What is a mixture?

26. List the four groups of organic compounds found in living things.

27. A chemical bond formed by the transfer of electrons is a(n) _____ bond.

28. A substance that speeds up the rate of a chemical reaction is called a(n) _____.

29. Starches and sugars are examples of what group of organic compounds?

    a) lipids
    b) nucleic acids
    c) proteins
    d) carbohydrates

30. Which one of the following is/are function(s) of nucleic acids in living things?

a) to store energy
b) to store and transmit genetic information
c) to transport substances into cells
d) to serve as chemical messengers

EXPLORING LIVING THINGS

1. A child cutting teeth is an example of _____.

   a) development
   b) growth
   c) respiration
   d) adaptation

2. A shining light that causes you to shut your eyes is a _____.

   a) response
   b) need
   c) stimulus
   d) variable

3. Living organisms are made up of about 70 percent _____.

   a) water
   b) oxygen
   c) carbon dioxide
   d) minerals

4. The inherited characteristics that may help an organism live are called _____.

   a) stimuli
   b) adaptations
   c) minerals
   d) theories

5. _____ disproved the theory of spontaneous generation.

   a) Oparin   b) Pasteur
   c) Spallanzani   d) Redi

6. Which of the following is not a characteristic of all living things?

   a) ability to sexually reproduce
   b) ability to maintain a stable internal environment
   c) ability to move in response to the environment
   d) ability to grow and develop

7. The process that helps organisms keep their internal condition constant is called _____.

   a) evolution
   b) homeostasis
   c) metabolism
   d) photosynthesis

8. Biology is the study of _____.

    a) the land, water, and the air on Earth
    b) living things
    c) animals and plants only
    d) the environment

9. All of the following are characteristics of all living organisms except _____.

    a) growth
    b) reproduction
    c) movement
    d) the use of energy

10. Living organisms use _____ as their main source of energy.

11. What are the five characteristics that all living organisms share?

12. What is homeostasis?

13. What is metabolism?

14. The smallest units that carry on the functions of life are _____.

15. The _____ is the length of time an organism is expected to live.

16. Individual living things are called _____.

17. All the changes living organisms undergo as they grow are called _____.

18. A(n) _____ is a characteristic an organism has that makes it better able to survive in its   environment.

19. Explain why living organisms need energy.

20. Explain how a pigeon has all the characteristics of living organisms.

CELL STRUCTURE & FUNCTION

1. The cell structure that contains the cell's genetic material and controls many of the cell's activities is called the _____.

   a) nucleus
   b) organelle
   c) cell envelope
   d) cytoplasm

2. The work of Schleiden and Schwann can be summarized by the statement that _____.

   a) animals are made of cells
   b) plants are made of cells
   c) plants and animals are made of cells
   d) all plants and animals are made of cells

3. Eukaryotes generally contain _____.

   a) specialized organelles
   b) nucleus
   c) genetic material
   d) all of the above

4. Which of the following is not found in the nucleus?

   a) cytoplasm
   b) nucleolus
   c) chromatin
   d) DNA

5. Which organelle breaks down compounds into small particles that the cell can use?

a) lysosome   b) Golgi apparatus
c) endoplasmic reticulum   d) mitochondrion

6. Which organelle makes protein using coded instructions that come from the nucleus?

a) Golgi apparatus   b) mitochondrion
c) vacuole   d) ribosome

7. Which of the following is a function of the cell membrane?

a) to store water, salt, proteins, and carbohydrates
b) to break down lipids, carbohydrates, and proteins from foods
c) to keep the cell wall in place
d) to regulate which materials enter and leave the cell

8. An organ system is a group of organs that_____.

a) work together to perform a specific function
b) are made up of similar tissue
c) are made up of similar cells
d) work together to perform all the functions in a multi-cellular organism

9. A group of similar cells that performs a particular function is called a(n) _____.

    a) organ
    b) tissue
    c) organ system
    d) division of labor

10. Who was the first person to identify and see cells?

    a) Robert Hooke
    b) Anton van Leeuwenhoek
    c) Matthias Schleiden
    d) Rudolf Virchow

11. The thin, flexible barrier around a cell is called a _____.

    a) cell wall
    b) cell membrane
    c) cell envelope
    d) cytoplasm

12. Prokaryotes lack the following: _____.

    a) cell membrane
    b) nucleus
    c) cytoplasm
    d) genetic material

13. Which organisms contain a nucleus?

    a) prokaryotes
    b) eukaryotes
    c) bacteria
    d) organelle

14. The main function of the cell wall is to _____.

    a) support and protect the cell
    b) store DNA
    c) help the cell move
    d) direct the activities of the cell

15. Which of the following is a function of the nucleus?

    a) to control most of the cell's processes
    b) to store DNA
    c) to continue the information needed to make proteins
    d) all of the above

16. Draw and label the plant and animal cells.

(See Appendix A)

17. Eukaryotes contain structures that act as if they are
    specialized organs. These structures are called _____.

18. Molecules tend to move from an area where they are more concentrated to an area where they are less concentrated. This process is called _____.

19. What are two functions of the nucleus?

20. How do prokaryotes and eukaryotes differ?

21. According to the cell theory, new cells are produced from existing _____.

22. During cell division, chromatin condenses to form _____, which are threadlike structures containing genetic information.

23. The cell takes in food and water and eliminates waste through the _____.

24. What are the functions of the cell membrane?

25. In what organisms are cell walls found?

26. What is the main function of the cell wall?

27. What are mitochondria?

28. Where are chloroplasts found?

29. Write the functions of the following: ribosomes, endo-
    plasmic reticulum, Golgi apparatus, lysosome, vacuole.

CELL PROCESSES, CELL GROWTH & DIVISION

1. An atom's nucleus contains _____.

    a) protons and electrons
    b) neutrons only
    c) neutrons, electrons, and proteins
    d) protons and neutrons

2. Bacteria are absorbed into cells by _____.

    a) endocytosis
    b) diffusion
    c) exocytosis
    d) osmosis

3. The organic compounds in the chromosomes are called _____.

    a) lipids
    b) carbohydrates
    c) nucleic acids
    d) water molecules

4. If the movement of molecules requires energy, it is called _____.

    a) diffusion   b) active transport
    c) osmosis   d) passive transport

5. Producers use light energy to make _____.

    a) chlorophyll   b) glucose

    c) protein   d) starch

6. What are the organisms that can't make their own food?

    a) consumers

    b) enzymes

    c) producers

    d) organisms

7. _____ is a process that releases energy when oxygen is insufficient.

    a) equilibrium

    b) cellular respiration

    c) fermentation

    d) metabolization

8. Most of any cell's life is spent in a period of growth and development called the _____.

    a) metaphase

    b) interphase

    c) prophase

    d) telophase

9. All of the following are true of animal and plant cells during mitosis except _____.

    a) only animals have spindle fibers
    b) only plants have rigid cell walls
    c) only plants form cell plates
    d) only animals have centrioles

10. Each human skin cell has ____ pairs of chromosomes.

    a) 18   b) 13
    c) 23   d) 46

11. In sexual reproduction, a new organism is produced when _____.

    a) sex cells from two parents combine
    b) cells divide by mitosis
    c) an organism divides into two equal parts
    d) a new organism grows from the body of its parent

12. In ____, a new organism grows from just a part of the parent organism.

    a) fission
    b) mitosis
    c) regeneration
    d) sexual union

13. Human gametes have ____ individual chromosomes.

    a) 13
    b) 23
    c) 33
    d) 46

14. The number of chromosomes in a gamete of an organism is its ____ chromosome number.

    a) haploid
    b) diploid
    c) RNA
    d) zygote

15. Meiosis consists of ____ division(s) of the nucleus.

    a) one
    b) two
    c) three
    d) four

16. At the end of meiosis, ____ cells are the result.

    a) three
    b) two
    c) four
    d) five

17. In DNA, adenine always pairs with _____.

    a) cytosine
    b) guanine
    c) thymine
    d) uracil

18. Proteins are made up of units of _____ linked together in a specific order.

    a) amino acids
    b) centrioles
    c) centromeres
    d) cytoplasm

19. The code for making proteins is carried to the ribosomes by _____.

    a) cytosine
    b) DNA
    c) RNA
    d) thymine

20. The first phase of mitosis is called _____.

    a) anaphase
    b) prophase
    c) metaphase
    d) interphase

21. In which phase of mitosis do chromosomes become visible?

a) interphase
b) prophase
c) metaphase
d) telophase

22. What happens when cells come into contact with other cells?

a) they divide more quickly
b) they stop growing
c) they produce cyclins
d) they produce tumors

23. One of the principal compounds that cells use to store and release energy is called ___.

a) ATP
b) CO_2
c) O_2
d) NADPH

24. What is the principal pigment of plants?

a) CO_2
b) thylakoid
c) chlorophyll
d) sugar

25. The diffusion of water through a selectively permeable membrane is called _____.

26. Organisms that obtain energy from the foods they consume are called _____.

27. In the process of cell division, the division of the cell nucleus is called _____.

28. What is ATP and how is energy released from it?

29. Write the equation for photosynthesis in both symbols and words.

30. Write the equation for cellular respiration in symbols and words.

31. What are the four phases of mitosis?

32. Each pair of chromatids is attached to an area called the _____.

33. What is the cell cycle?

34. The four phases of mitosis.   (illustration)

   (See Appendix A)

35. A tiny structure located in the cytoplasm near the nuclear envelope is a(n) _____.

36. A fan-like microtubule structure that helps separate the chromosomes is a(n) _____.

37. The chromosomes in the nucleus of a cell contain a chemical code called _____.

38. The part of the chromosome that directs the making of a specific protein is a(n) ____.

39. Any permanent change in a gene or a chromosome is a(n) _____.

40. What do spindle fibers do during mitosis and meiosis?

41. If one strand of DNA had bases ordered ATCCGTC, what would be the bases of its other strand?

42. RNA differs from DNA in that it contains _____.

43. Chromosomes are doubled during _____.

44. What is cancer?

45. The division of a cell's cytoplasm is called _____.

46. The division of the cell nucleus is called _____.

47. The final phase of mitosis is called _____.

48. A cell becomes larger because _____.

    a) the surface area increases faster than its surface area
    b) volume increases faster than its surface area
    c) volume increases, but its surface area stays the same
    d) surface area stays the same, but its volume stays the same

49. The following can be problems when a cell grows EXCEPT
    _____.

    a) excess oxygen
    b) DNA overload
    c) obtaining enough food
    d) expelling waste

50. Which of the following is NOT a way that cell division
    solves the problem of cell growth?

    a) cell division provides each daughter cell with its own
       copy of DNA
    b) cell division increases the mass of the original cell
    c) cell division increases the surface area of the original
       cell
    d) cell division reduces the original cell's volume

51. When are chromosomes visible during the cell cycle?

    a) only when they are being replicated
    b) only during interphase
    c) only during metaphase
    d) only during G1 phase

52. When is a cell's DNA replicated during the cell cycle?

    a) in the G1 phase
    b) in the G2 phase
    c) in the S phase
    d) in the M phase

53. Which event occurs during interphase?

    a) the cell grows
    b) spindle fibers begin to form
    c) centrioles appear
    d) centromeres divide

54. During which phase of mitosis do the chromosomes line up along the middle of the dividing cell?

    a) prophase
    b) telophase
    c) metaphase
    d) anaphase

55. Which of the following represents the phases of mitosis in their proper sequence?

    a) interphase, prophase, metaphase, anaphase, telo-
       phase
    b) prophase, metaphase, anaphase, telophase
    c) interphase, prophase, metaphase, telophase
    d) prophase, metaphase, anaphase, telophase, cytokine-
       sis

56. What is the role of the spindle during mitosis?

    a) it breaks down the nuclear membrane
    b) it helps separate the chromosomes
    c) it duplicates the DNA
    d) it divides the cell in half

57. The two main stages of cell division are called _____.

    a) mitosis and interphase
    b) synthesis and cytokinesis
    c) mitosis and cytokinesis
    d) the M and S phase

58. Cancer is a disorder in which some cells have lost the ability to control their _____.

    a) spindle fibers
    b) size
    c) growth rate
    d) surface area

59. What is cell division?

60. Describe what happens during the four phases of mitosis.

61. Label the four phases of mitosis.

    (See Appendix A)

62. Draw, label, and color pictures of the stages of mitosis.

63. List two problems that growth causes for cells.

64. Why are chromosomes not visible in most cells except during cell division?

65. What organelles in cells break down food molecules and release energy?

66. The ____ structure forms the outer boundary of the cell.

67. ____ are responsible for the packaging and secreting substances within the organelles of the cell.

68. What is the gel-like material outside the nucleus and inside the cell membrane?

HEREDITY & GENETICS

1. The study of how alleles affect offspring is called _____.

2. The passing of traits from one generation to another is called ____.

3. The genetic makeup of an organism is called a _____.

4. A trait that is hidden is called _____.

5. The different forms of the same gene are called _____.

6. An allele inherited on a sex chromosome is called a _____.

7. What are gametes?

8. The likelihood that a particular event will occur is called _____.

9. How do geneticists use Punnett squares?

10. Why can the principles of probability be used to predict the outcomes of genetic   crosses?

11. Complete the Punnett square to show the possible gene combinations for the offspring.

Punnett Square for Tt x Tt

	T	t
T		
t		

12. What does it mean when two sets of chromosomes are homologous?

13. Blood type in humans is controlled by ____ alleles.

14. What happens when a group of gene pairs act together to control a single trait?

15. The pink four o'clock flowers in Mr.Mendel's experiment occurred as a result of ____.

16. What pair of sex chromosomes is in each of your cells?

17. What are the chances of a color-blind man passing the trait on to his son?

18. What are the chances that the hemophilia gene will be passed from a woman carrier to   any of her sons?

19. What is the difference between homozygous and heterozygous?

20. Compare and contrast multiple allele inheritance and polygenic inheritance.

21. Explain the importance of Mr.Mendel's experimental methods.

22. Using an example, explain how blood typing could be used to detect the correct parentof a newborn if there was a mix-up in a nursery.

GENETICS GUIDE QUESTIONS

1. What is located on chromosomes?

   a) genes
   b) carbohydrates
   c) pedigrees
   d) zygotes

2. Color blindness results from an alike allele that is ____.

   a) dominant
   b) on the X chromosome
   c) the Y chromosome
   d) present only in females

3. During meiosis, ____ for a trait separate.

   a) proteins
   b) alleles
   c) cells
   d) chromosomes

4. The major function of genes is to control ____.

   a) traits
   b) chromosomes
   c) cell membranes
   d) mitosis

5. Sickle-cell anemia is inherited through _____.

    a) polygenic inheritance
    b) recessive genes
    c) multiple alleles
    d) incomplete dominance

6. Eye color is a trait inherited through _____.

    a) polygenic inheritance
    b) multiple allele
    c) incomplete dominance
    d) recessive genes

7. A female is produced if an egg unites with a sperm containing _____.

    a) XX chromosomes
    b) X chromosomes
    c) Y chromosomes
    d) XY chromosomes

8. Punnett squares are used to _____ the outcome of crosses of traits.

    a) predict
    b) dominate
    c) number
    d) analyze

9. Gregor Mendel used pea plants to study ____.

    a) flowering
    b) the inheritance of traits
    c) gamete formation
    d) cross-pollination

10. Mendel's "factors" are now called ____.

    a) traits
    b) alleles
    c) genes
    d) characters

11. Two plants with the genotypes TT and Tt ____.

    a) would have the same phenotypes
    b) would have different phenotypes
    c) have all dominant alleles
    d) have all recessive alleles

12. The principle of dominance states that _____.

    a) all alleles are recessive
    b) all alleles are dominant
    c) some alleles are dominant and others are recessive
    d) alleles are neither dominant nor recessive

13. Organisms that have two identical alleles for a particu-
    lar trait are said to be _____.

    a) heterozygous
    b) hybrid
    c) homozygous
    d) dominant

14. The situations in which one allele for a gene is not com-
    pletely dominant over another allele for that gene are
    called _____.

    a) multiple alleles
    b) incomplete dominance
    c) dominant alleles
    d) multiple genes

15. Mr.Mendel's principles of genetics apply to _____.

    a) animals only
    b) plants only
    c) pea plant study
    d) all organisms

16. Gametes are produced by the process of _____.

    a) meiosis
    b) mitosis
    c) crossing over
    d) replication

17. Gene maps are based on _____.

    a) independent assortment
    b) frequencies of crossing over between genes
    c) genetic diversity
    d) the number of genes in a cell

18. If an organism's diploid number is 12, its haploid number is _____.

    a) 12
    b) 6
    c) 24
    d) 3

19. The principles of probability can be used to _____.

    a) predict the traits of the offspring produced by genetic crosses
    b) predict the traits of the parents used in genetic crosses
    c) determine the actual outcomes of genetic crosses
    d) decide which organisms are the best to use in genetic crosses.

20. Mr.Mendel concluded that traits are _____.

    a) not inherited through the passing of factors from parent to offspring

    b) inherited through the passing of factors from parents to offspring

    c) determined by dominant factors only

    d) determined by recessive factors only

21. How many sets of chromosomes are in a diploid cell?

22. What happens to the number of chromosomes per cell during meiosis?

23. The person described as the "father of genetics" is

    _____.

    a) Mendel

    b) Darwin

    c) Genome

    d) Lamarck

24. A ____ is a tool used for tracing occurrences of a trait in a family.

    a) Punnett square

    b) gene map

    c) pedigree

    d) gene pool

25. Body parts that are similar in origin and structure are called _____.

    a) heterozygous
    b) homologous
    c) homozygous
    d) heterologous

26. Height is controlled by _____.

    a) multiple alleles
    b) polygenic inheritance
    c) sex-linked alleles
    d) a single gene

27. Sickle-cell anemia is an example of _____.

    a) a genetic disorder
    b) a sex-linked disorder
    c) incomplete dominance
    d) polygenic inheritance

28. Hemophilia affects blood _____ in the human body.

    a) circulation
    b) clotting
    c) oxygen
    d) pressure

29. Genetic engineering has already helped many people by
_____.

    a) curing diseases
    b) altering pedigrees
    c) producing medicine
    d) eliminating infant death

30. Blood type is a product of _____.

    a) multiple alleles
    b) a pair of genes
    c) polygenic inheritance
    d) sex-linked genes

EVOLUTION

1. Explain how genetics and the theory of evolution are related.

2. Explain what is meant by the term "survival of the fittest."

3. Explain what would likely happen to an albino deer in its natural environment.

4. A group of organisms whose members look alike and successfully reproduce among themselves is called a _____.

5. Darwin concluded that individuals with traits most favorable for a specific environment survive and pass these traits on to their offspring.This theory of evolution is called _____.

6. A specific group of organisms of some species that live in an area is called a _____.

7. Compare Lamarck's explanation of evolution with Darwin's theory.

8. The remains of life from an earlier time (ex. a million years ago) are called _____.

9. Most fossils are found in _____.

10. Fossils are a record of organisms that lived in the _____.

11. Body parts that are similar in origin and structure are called ____ structures.

12. List five examples of evidence that support the theory of evolution.

13. Why is DNA considered evidence that monkeys, apes, and humans all evolved from a common ancestor?

14. Explain at least three kinds of evidence that suggests all primates shared a common ancestor.

15. What is the importance of Australopithecus?

16. Which order do the monkeys, apes, and humans belong to?

17. Mass extinctions have occurred several times in Earth's history. Why are scientists so concerned about the extinction of species today?

18. Use Darwin's theory of natural selection to explain how the giraffe got its long neck.

19. Describe the process a scientist would use to figure out the age of a fossil.

20. Homologous structures, vestigial structures, and fossils all provide evidence of ____.

a) extinction
b) species population
c) food choice
d) evolution

21. The most accurate age of a fossil can be found using

    ____.

    a) natural selection
    b) radioactive elements
    c) relative dating
    d) camouflage

22. A factor that controls natural selection is/are _____.

    a) unused traits that become smaller
    b) inheritance of acquired traits
    c) organisms producing more offspring than can survive
    d) the size of the organism

23. A series of helpful variations in a species may result in

    ____.

    a) adaptation
    b) fossils
    c) extinction
    d) climate change

24. Organisms adapting to their environment are _____.

    a) extinct
    b) surviving and reproducing
    c) not producing
    d) forming fossils

25. The red lion and Siberian tiger are examples of _____.

    a) fossils
    b) hominids
    c) extinct species
    d) endangered species

26. The earliest known drawings were done by ____.

    a) Cro-Magnon humans
    b) Australopithecus
    c) Homosapiens
    d) Homo habilis

27. Darwin's theory of evolution is based on the idea of ____.

    a) use and disuse
    b) natural variation and natural selection
    c) tendency toward perfect, unchanging species
    d) the transmission of acquired characteristics

28. The idea that only famine, disease, and war could prevent the endless growth of the human population was presented by ____.

    a) Jean-Baptiste Lamarck
    b) Thomas Malthus
    c) Charles Darwin
    d) Charles Lyell

29. The scientist who attempted to explain how rock layers form and change over time was

    a) James Hutton
    b) Thomas Malthus
    c) Charles Darwin
    d) Jean-Baptiste Lamarck

CLASSIFICATION OF LIVING ORGANISMS

1. What is classification?

2. Characteristics used to classify living things include the cell ____.

3. One of the characteristics of the plant kingdom is that ____.

4. List the four kingdoms of life.

5. Explain the Linnaean system of classification.

6. What is taxonomy?

7. Explain the difference between the animal kingdom and the fungi kingdom.

8. The scientific first name of the organism is its ____.

9. Fungi always _____.

10. Bacteria include _____.

   a) the cells in the environment and our bodies that per-form many functions
   b) contains cells without nucleus
   c) eukaryotes cells
   d) move independently

11. The second part of a scientific name is different from the _____.

   a) genus species
   b) family genus
   c) order of family
   d) genus in family

12. Several kinds of classes make up a ____.

    a) phylum
    b) kingdom
    c) genus in family
    d) family in order

13. Two kingdoms that Mr. Linnaeus recognized were ____.

    a) fungi and plants   b) animals and bacteria
    c) animals and plants   d) protists and fungi

14. In the six kingdoms of classification, which kingdom was part of the plants group?

    a) carnivores   b) animalia
    c) protists and plants   d) fungi and plants

15. Plants always ____.

    a) make their own food
    b) have cells without membranes
    c) move independently
    d) all of the above

16. The kingdom that includes more celled and one celled organisms is called ____.

    a) fungi   b) plant
    c) animal   d) bacteria

17. The group called "organisms" includes _____.

    a) only small and specific categories
    b) one large category of organisms
    c) categories of living things
    d) large and small categories of living things

18. The warm-blooded animals that have body hair and produce milk are part of the _____.

    a) reptilian group
    b) amphibian group
    c) mammalian group
    d) kingdom group

BACTERIA & VIRUSES

1. What are bacteria?

2. What are viruses?

3. What do all viruses possess?

4. What happens when viruses enter the inside of a cell?

5. What are the two ways the human body protects itself against viruses?

6. What are pathogens?

7. What are antibiotics?

8. List two ways that bacterial cells differ from animal cells.

9. List a variety of ways in which communicable diseases can be spread.

10. The reason tuberculosis returns is because the bacteria that causes it _____.

    a) does not spread easily from an infected organism
    b) develops resistance against antibiotics
    c) can live without energy
    d) helps to clean up soil

11. Bacteria does not have ____.

    a) chlorophyll
    b) a cell wall
    c) nuclear material
    d) a Golgi apparatus

12. Bacteria reproduces by _____.

    a) fission
    b) budding
    c) mating
    d) fertilizer

13. The following are all true of bacteria except: _____.

    a) they are prokaryotic cells
    b) they are eukaryotic
    c) they have cell walls
    d) they have cell membranes

14. Organisms whose cells possess no membrane are called
_____.

a) eukaryotes
b) prokaryotes
c) fissions
d) budding

15. Bacteria possess a thick ____ that surrounds the cell
wall.

a) cell membrane
b) chlorophyll
c) capsule
d) nucleus

16. Bacteria have a whip-like tail called a ____.

a) flagellum
b) capsule
c) paramecium
d) cilia

17. Which of the following describes the role of bacteria in
our environment?

a) recycling nutrients
b) carrying out photosynthesis
c) nitrogen fixing
d) all of the above

18. Which one of the following diseases is NOT caused by a virus?

a) AIDS/HIV
b) malaria
c) tetanus
d) chicken pox

19. What does a vaccine do when it is injected into the human body?

a) it produces toxins that change the bacterial balance
b) it causes the body to produce immunity to fight disease
c) it destroys pathogens
d) it causes disease in humans

20. Which one of the following diseases is caused by a virus?

a) strep throat
b) malaria
c) foot decay
d) AIDS

21. What are viruses composed of?

a) DNA or RNA surrounded by a protein cell
b) cell membrane
c) protein
d) nitrogen fixers

22. Which one of the following describes prokaryote cells?

    a) prokaryote movement
    b) cell shape
    c) the variety of ways prokaryotes maintain energy
    d) all of the above

23. Draw, label, and illustrate a bacterial cell.

    (See Appendix A)

PLANTS

1. List three methods by which plant seeds are dispersed.

2. What is the common pigment in most plants called?

3. What is the chemical formula for sugar or glucose?

4. What is an ethnobotanist?

5. What is a plant?

6. What tissue is responsible for moving food from leaves to other parts of the plant?

7. What tissue is responsible for producing new xylem and phloem cells?

8. A plant's male reproductive organ is called a _____.

9. A plant's female reproductive organ is called a _____.

10. The sticky area where pollen grains land is called a
_____.

11. The pores in a plant's leaf surface are called _____.

12. What happens during the process of photosynthesis?

13. The roots move water and _____ from the soil through
the plant stems to the leaves.

14. The trunk of a tree is called _____.

15. List the four basic needs of plants.

16. The transfer of pollen grains from the male to the
female reproductive cell is called ____.

17. List the three structural organs of seed plants.

18. What is a fruit?

19. What is the overall equation for photosynthesis?

20. What are the functions of plant roots?

21. List the functions of the stem.

22. Explain how we can find out the age of a tree.

23. What is transpiration?

24. The outer layer of a seed that protects the embryo and nutrient supply is called ____.

WORMS & MOLLUSKS

1. List the three body parts of a mollusk.

2. What is the thin layer of tissue that covers the mollusk's body?

3. Describe how the mollusk shell is formed.

4. Explain how land snails and slugs live only in moist environments.

5. Roundworms exchange oxygen and excrete metabolic waste from their _____.

6. How do roundworms reproduce?

7. Explain where hookworm eggs hatch and develop.

8. The head of a mature tapeworm is called a _____.

9. The tapeworm uses its scolex to _____.

10. What is a hermaphrodite worm?

11. What usually happens during fission?

12. Explain how earthworms improve soil.

13. How many chambers are in the earthworm heart?

14. Name the organ in earthworms that functions as the digestive system and grinds food into small pieces.

15. Draw, illustrate, and label an earthworm.

   (See Appendix A)

FISH, AMPHIBIANS & REPTILES

1. Explain how fish exchange gases.

2. Explain the function of the fish bladder.

3. List two ways that fish fins are responsible for movement.

4. What is an amphibian?

5. State two ways in which amphibians protect themselves from enemies or predators.

6. List three characteristics of reptiles.

7. What is a reptile?

8. Compare and contrast the lungs of amphibians and reptiles.

9. List the three groups of modern amphibians.

10. Explain the disadvantage of reptile skin.

11. The structures that fish use to obtain oxygen from water are called ____.

a) gills
b) lungs
c) scales
d) nerve cord

12. Fish tend to get rid of nitrogen waste through _____.

a) eliminating ammonia from the gills and kidney
b) eliminating gases from the gills
c) taking in ammonia gases through the gills
d) taking in water through the lungs

13. Explain why amphibians are dependent on water.

14. Which features show the differences between fish and amphibians?

a) scales
b) the vertebral
c) breathing through the gills
d) live part of their lives in water

15. The salamanders that live on land are missing which of the following?

    a) kidneys
    b) lungs
    c) scales
    d) legs

16. Which is NOT a characteristic of the amphibian circulatory system?

    a) a heart with five chambers
    b) strong bones in the pelvic area
    c) the separation of oxygen-rich and less oxygen-rich blood
    d) the shape of their bodies

17. Which organ(s) adjust(s) the buoyancy of bony fish?

    a) kidney   b) swim bladder
    c) scales   d) gills

18. Which one of the following are ectotherms that spend their time in water and on land, and lay their eggs in water?

    a) reptiles   b) amphibians
    c) bony fish   d) salamanders

19. Which one of the following are ectotherms with dry, scaly skin and that lay amniotic eggs?

    a) fish
    b) reptiles
    c) salamanders
    d) amphibians

20. Turtles breathe oxygen through their _____.

    a) lungs
    b) skins
    c) gill slits
    d) nostrils

21. Most reptile hearts contain _____.

    a) one chamber
    b) two chambers
    c) three chambers
    d) four chambers

22. Amphibian eggs dry out quickly because _____.

    a) they do not have shells
    b) they are fertilized outside
    c) they are usually laid in water
    d) they live in water

23. Most reptiles reproduce through _____

    a) internal fertilization, and they are viviparous
    b) internal fertilization, and they are oviparous
    c) external fertilization
    d) external fertilization and water temperature

24. Which of these organisms contains a four-chamber heart?

    a) snakes   b) turtles
    c) alligators   d) fish

25. Reptiles can live in every climate except _____.

    a) dry climates   b) wet climates
    c) hot climates   d) cold climates

26. An endotherm is an organism that _____.

    a) stores a small amount of food
    b) lives in water
    c) generates its own body heat
    d) maintains a low temperature

27. Which one of the following expresses how endotherms get rid of heat?

   a) sweating or panting
   b) lying in the sun
   c) swimming in water
   d) moving around during cold nights

ECOSYSTEMS

1. List four ecosystems in the community where you reside.

2. Explain why palm trees cannot survive in a tundra area.

3. What is a succession?

4. What kinds of organisms live in an older pond's shallow water?

5. What is an ecosystem?

6. Explain what a greenhouse gas is.

7. What is an ecological resource?

8. Describe the duties that pioneer species play in a primary succession community.

9. Lakes and ponds include ____.

    a) deserts   b) ecosystems
    c) standing water ecosystems   d) biomes

10. One way a primary succession community begins is with ____.

    a) climate change   b) community
    c) lava flow   d) population growth

11. Which one of the following features the difference between primary and secondary succession?

    a) secondary succession begins on soil and primary succession begins on new surface
    b) primary succession begins in a place that has never supported living organisms
    c) secondary succession begins in a place that was once the home of living organisms
    d) secondary and primary succession are exposed to new environments

12. Which of the following contributes to the continual change in the ecosystem?

a) climate changes in the environment
b) more disturbances
c) new non-native species
d) all of the above

13. The following determine climate except _____.

a) rivers   b) sunlight
c) rainwater   d) wind temperatures

14. Grassland biomes include the following except _____.

a) wheat fields
b) occasional fires
c) plants for grazing organisms
d) temperature ranges

15. Which one of the following organisms does not need sunlight?

a) large trees   b) chemosynthetic bacteria
c) algae bacteria   d) soil depletion

16. An organism that uses energy to produce its own food from inorganic compounds is called a(n) _____.

a) consumer   b) heterotroph
c) autotroph   d) species

17. Animals that eat both producers and consumers are called _____.

    a) omnivores   b) herbivores

    c) third-level consumers   d) second-level consumers

18. A snake that eats a rat that has eaten an insect that has eaten grass is called a _____.

    a) first-level consumer   b) first-level producer

    c) third-level consumer   d) second-level producer

19. A green planet includes _____.

    a) consumers   b) producers

    c) herbivores   d) carbohydrates

20. The members of a particular species that reside in the same community are called a ___.

    a) community   b) ecosystem

    c) biome   d) population

21. The original source of all the energy in an ecosystem is _____.

    a) water   b) sunlight

    c) CO_2 d) the food chain

22. The branch of biology dealing with the connections among organisms and their environment is called _____.

a) ecology   b) natural resources

c) omnivore   d) the food web

23. Organisms that cannot produce their own food are called _____.

a) producers   b) heterotrophs

c) omnivores   d) decomposers

24. An organism that depends on plants for food is called a(n) _____.

a) carnivore   b) omnivore

c) herbivore   d) consumer

25. What is the difference between a food web and a food chain?

26. Plant-eating organisms such as goats are called _____.

27. Draw a diagram of a food web.

(See Appendix A)

28. What is commensalism?

29. An example of a biotic factor is a ____.

    a) consumer   b) producer
    c) predator   d) prey

30. The place where an organism resides in a community is called a _____

    a) habitat   b) community succession
    c) niche   d) prey

31. The ecological succession that begins with the environment that does not have soil or support living things is called _____.

32. The first community of organisms to move to a new environment is called _____.

33. The community that reaches the final climax of ecological succession is called a _____.

THE SKELETAL SYSTEM

1. Describe the skeletal system.

2. List the functions of the skeletal system.

3. The cells that produce bones are called _____.

4. The two separate minerals that support the mass of bones are called _____.

5. The skeleton of an embryo is composed with connective tissue called _____.

6. What is a joint bone?

7. List the three types of muscle tissue.

8. Explain the functions of smooth muscles.

9. Describe the two major parts of the skeletal system.

10. Explain why regular exercise is important for the body.

11. List five major functions of the skin.

12. The human skeleton contains _____ bones of various shapes and sizes.

    a) 160   b) 106

    c) 206   d) 250

13. What makes bones hard?

    a) proteins   b) minerals

    c) fat   d) carbohydrates

14. The muscles in the body include _____ muscles.

    a) smooth   b) involuntary

    c) cardiac   d) skeletal

15. The major functions of the skull include _____.

    a) protecting the heart and organs

    b) producing blood cells

    c) protecting the skeleton

    d) protecting the brain

16. Which one of these features provides support for the body and protection for internal organs?

    a) muscles and bones   b) joints
    c) skin   d) skeleton

17. The skeleton of a newborn baby is composed of _____.

    a) bone cells   b) skeleton
    c) cartilage   d) ligaments

18. The two layers that make up the skin include _____.

    a) epidermis and melanin   b) epidermis and dermis
    c) hair and follicles   d) keratin and epidermis

19. What percentage of the human body is made up of muscle?

    a) 20 percent   b) 10 percent
    c) 25 percent   d) more than 50 percent

20. Which one of the following is NOT a function of the skin?

    a) removing body waste
    b) regulating body temperature
    c) relaxing the muscles
    d) protecting the body from disease

21. Fingers, elbows, and knees all have a _____.

  a) ball and socket   b) hinge
  c) pivot   d) joint

NUTRITION & DIGESTIVE SYSTEM

1. What are antioxidants?

2. Explain the benefits of fruits and vegetables.

3. What kinds of foods are good sources of antioxidants?

4. The teeth begin to break dow during _____ digestion.

5. What major organ structures make up the digestive system?

6. Explain how the body loses water.

7. The energy saved in food is measured in units called _____.

8. What is nutrition?

9. List five nutrients that the body requires.

10. Explain the functions of food.

11. What is the function of the digestive system?

12. Tell where the most chemical digestion takes place.

13. Explain the liver's role in digestion.

14. What is the primary duty of the large intestine?

15. Many cultures have used ____ to meet their needs for protein as well as for carbohydrates.

    a) vegetables   b) whole grains
    c) meat   d) fish

16. The process of digestion starts at your _____.

    a) mouth   b) stomach
    c) enzymes   d) small intestine

17. The essential, organic nutrients that help your body use other nutrients are called ____.

    a) proteins   b) vitamins

    c) calcium   d) fats

18. The saturated fats found in ____ contribute to high cholesterol and heart disease.

    a) fish   b) red meats

    c) eggs   d) fats

19. The breakdown of food so that it can be taken into the cells is called ____.

    a) digestion   b) peristalsis

    c) saliva   d) chime

20. Chewing food in the mouth is an example of ____.

    a) chemical digestion   b) mechanical digestion

    c) small intestine   d) enzyme

21. How many food groups make up the food pyramid?

    a) six   b) five

    c) four   d) three

22. What materials does the body need for growth and repair?

   a) carbohydrates   b) proteins
   c) water   d) fats

23. What substances are needed for body growth and repair?

   a) nutrients   b) enzymes
   c) water   d) vitamins

24. Which vitamin is very important for good eyesight?

   a) vitamin A   b) vitamin C
   c) vitamin D   d) vitamin E

25. Carbohydrates are in foods such as pasta and _____.

   a) grains   b) fruits
   c) poultry   d) meat

CIRCULATORY SYSTEM

1. List the parts of the circulatory system.

2. The smallest blood vessels we have discovered in the body are called _____.

3. Name three ways to avoid cardiovascular diseases.

4. Explain what blood pressure is in the human body.

5. The blood vessels that carry blood toward the heart are _____.

6. The flow of blood to the heart, lungs, and back to the heart is called _____.

7. A stroke can happen if an artery in the ____ is clogged up.

a) lungs   b) brain
c) heart   d) kidney

8. During pulmonary circulation, blood passes through two organs called the _____.

    a) heart and lungs   b) brain and lungs
    c) lungs and heart   d) heart and oxygen

9. High blood pressure is also called _____.

    a) hypertension   b) leukemia
    c) heart murmur   d) heart disease

10. Red blood cells are produced in the _____.

    a) muscle   b) white blood cells
    c) bone marrow   d) plasma

11. Which one of the body systems acts as a transportation system?

    a) excretory system   b) circulatory system
    c) respiratory system   d) digestive system

12. Which one of the features are the smallest blood vessels?

    a) capillaries   b) veins
    c) arteries   d) valves

13. The blood cells that contain hemoglobin are called _____.

    a) white blood cells   b) red blood cells
    c) platelets   d) blood vessels

14. Which one of these is not a part of the circulatory system?

    a) blood vessels   b) heart
    c) kidney   d) air passage

15. Which one of the following serves as the large circulatory system?

    a) pulmonary circulation   b) systemic circulation
    c) coronary circulation   d) circulatory system

16. Which one of the following describes the functions of the valves in the circulatory system?

    a) they prevent the backward flow of blood
    b) they carry oxygen from your lungs to your body
    c) they transport nutrients
    d) lower chamber of the heart

17. What is the circulatory system?

18. Draw and label the circulatory system.

See Appendix A

RESPIRATORY & EXCRETORY SYSTEM

1. List the functions of the urinary system.

2. The four organs that are used for excretion are the
   ____.

3. The process by which waste is removed is called ____.

4. What is the excretory system?

5. The removal of waste from blood with the use of a
   machine is called ____.

6. The primary organ of excretion is called a ____.

7. The excretory sac-like organ that stores urine is called
   ____.

8. What is the function of the ureter?

9. Explain how the body loses water.

10. During excretion, the kidneys are important in maintaining _____ in the body.

11. What is the function of the respiratory system?

12. The part of the brain that controls breathing is called the _____.

13. Explain why passengers in airplane emergency situations often have to be asked to begin breathing pressurized oxygen.

14. The movement of air into and out of the lungs is called _____.

15. The gas exchange happens in the _____.

16. What is respiration?

17. What usually prevents food from entering the trachea?

18. Explain what occurs when you inhale.

19. The smallest tubes in the lungs are called _____.

20. Explain why carbon monoxide is dangerous to your health.

21. What is emphysema?

22. List three respiratory diseases caused by smoking.

23. Inhaling the smoke of others is called _____.

24. Explain why it is so difficult to quit smoking.

25. Is the following statement true or false?

   Nicotine is a stimulant drug that increases pulse rate and blood pressure.

26. What are the functions of the trachea, bronchi, lungs, and alveoli?

27. Explain the directions your rib cage and diaphragm move as you breathe.

28. What causes a person to choke while eating food?

29. When a person breathes, the lungs take in oxygen and remove _____.

a) oxygen   b) nitrogen
c) carbon dioxide   d) hydrogen

30. Cell respiration involves supplying your body with _____.

a) nitrogen   b) oxygen
c) hydrogen   d) tar

31. Which one of the following prevents foods or liquids from passing through the   trachea?

a) larynx   b) pharynx
c) epiglottis   d) bronchi

32. Which one of the following is a tube-like passage for both food and oxygen?

a) trachea   b) epiglottis
c) larynx   d) pharynx

33. Inside the lungs, the exchange of oxygen and CO_2 happens between the _____.

a) diaphragm   b) alveoli
c) bronchi   d) pharynx

34. Which one of the following in cigarettes contributes to lung cancer?

a) nicotine   b) tar
c) CO_2 d) hydrogen

35. From the following, which one signifies that a person is breathing?

a) feeling air being exhaled from the nose
b) chest rising and falling
c) air being exhaled from the mouth
d) all of the above

36. Air is filtered, warmed, and moistened in the _____.

a) lungs   b) nose
c) diaphragm   d) heart

37. Which one of the following forces air into the lungs by contraction?

a) diaphragm   b) alveoli
c) heart   d) bronchi

38. Generally, what organ controls breathing?

a) diaphragm   b) heart
c) lungs   d) brain

39. Which one of the following is not contained in tobacco?

a) nicotine   b) tar
c) coffee   d) carbon monoxide

40. Which one of the following is not a disease caused by smoking?

a) emphysema   b) chronic bronchitis
c) lung cancer   d) pneumonia

41. The excretory organs of the body are the kidneys, lungs, and _____.

a) liver   b) kidney
c) skin   d) pancreas

42. Which one of the following is the urinary organ system?

a) respiratory   b) circulatory
c) excretory   d) digestive

43. Which one of the following is not the process of losing
     water from the body?

a) respiration   b) circulation
c) sweating   d) urinating

44. Waste passes into the rectum through the _____.

a) small intestine   b) large intestine
c) bladder   d) stomach

45. The function of the excretory system includes _____.

a) removing waste   b) circulating blood
c) breaking down food   d) dissolving food

46. Usually, perspiration includes water and _____.

a) blood   b) ammonia
c) salt   d) alcohol

47. Which one of the following is not an organ used for
     excretion?

a) kidney b) skin

c) lungs d) pharynx

48. Which one of the following do kidneys filter out of the blood?

a) blood b) excess salt

c) oxygen d) CO_2

49. Which one of the following are filtered out from the blood?

a) salts b) water

c) amino acids d) all of the above

50. The removal of waste from the blood with a machine is called _____.

a) dialysis b) meters

c) the pancreas d) a Bowman capsule

NERVOUS SYSTEM

1. What is the function of the nervous system?

2. Explain how the neurons are classified.

3. List the three types of neurons.

4. The central nervous system includes the ____ and the ____.

5. Define the function of the central nervous system.

6. Explain how a nerve impulse begins.

7. Explain the advantage of a reflex.

8. What is a reflex?

9. What is the nervous system?

10. What are sensory receptors?

11. List the four kinds of sensory receptors.

12. State what the autonomic nervous system regulates.

13. Explain how drugs disrupt the function of the nervous system.

14. Explain how alcohol affects the central nervous system.

15. Draw and label the parts of the neurons and nervous system.

16. What is the difference between dendrites and axons?

17. The space between one neuron and the next is known as a(n) _____.

a) synapse   b) axon
c) dendrite   d) reflex

18. The automatic response to stimuli is called a(n) _____.

a) neuron   b) synapse
c) reflex   d) axon

19. Which part of the neuron body cell receives messages?

    a) dendrite   b) interneuron
    c) nerve   d) synapse

20. Which one of the following sends impulses to the brain?

    a) sensory neurons   b) motor neurons
    c) reflexes   d) synapses

21. Which one of the following transmits impulses from the brain to muscles or glands?

    a) reflexes   b) sensory neurons
    c) dendrites   d) motor neurons

22. The nerve cells throughout the brain and spinal cord are called _____.

    a) dendrites   b) sensory neurons
    c) interneurons   d) reflexes

23. The large section of the brain that is divided into two parts is called the _____.

    a) brain stem   b) cerebellum
    c) cerebrum   d) midbrain

24. What part of the nervous system controls the reflex responses?

    a) cerebrum   b) midbrain

    c) spinal cord   d) dendrite

25. The process by which organ systems maintain a constant internal environment is called _____.

    a) blood circulation   b) homeostasis

    c) dendrite   d) photoreceptors

26. The function of the central nervous system is to _____.

    a) process information   b) relay information

    c) analyze information   d) all of the above

27. Drugs that raise the heart rate, blood pressure, and breathing rate are called ____.

    a) depressants   b) stimulants

    c) alcohol   d) marijuana

28. A group of similar cells that perform the same task is called _____.

    a) an organ   b) nerve cells

    c) tissue   d) the nervous system

29. The two divisions of the human nervous system are the central and the ____.

a) sensory receptors   b) endocrine
c) peripheral   d) nervous

30. Which of the following is a task of the cerebrum?

a) to control heart rate
b) to control activities of the body
c) to control breathing
d) to control pain receptors

31. The sense organs division includes the _____.

a) central nervous system   b) brain stem
c) peripheral nervous system   d) motor neuron

HUMAN REPRODUCTION

1. The joining of egg and sperm is called _____.

2. The canal that leads from the uterus to the outside of the body is called the _____.

3. After the female's egg is fertilized, it is called a _____.

4. The mother and embryo exchange oxygen, food, and waste products through the _____.

5. After eight weeks of development, the embryo is called a _____.

6. The female gonads are also called _____.

7. The testes are stored in a sac called the _____.

8. Explain why the testes are located outside rather than inside the body.

9. The semen consists of _____.

10. How often are eggs discharged from the uterus?

11. What is ovulation?

12. What is puberty?

13. What happens to the menstrual cycle when a female reaches menopause?

14. Why is human reproduction important?

15. How many chromosomes are in a male's sperm cell?

16. How many chromosomes does a zygote inherit from its parents?

17. What is the result when two eggs are released and fertilized by two different sperm?

18. The female eggs are produced in the ____.

    a) oviduct   b) ovaries
    c) vagina   d) uterus

19. The female egg is fertilized in the _____.

    a) uterus   b) oviduct
    c) womb   d) ovaries

20. Which one of the following glands controls the female menstrual cycle with hormones?

    a) prostate gland   b) thyroid gland
    c) pituitary gland   d) adrenal gland

21. What is the correct number of sperm released from the male during ejaculation?

    a) 20–30 million   b) 250–300 thousand
    c) 200–300 million   d) 10–15 million

22. The male's sperm and the female's egg each contain ____.

    a) 23 chromosomes   b) 21 chromosomes
    c) 27 chromosomes   d) 43 chromosomes

23. What results when an egg and sperm combine together?

    a) a fetus   b) an embryo

    c) a zygote   d) childhood

24. The onset of puberty normally begins between the ages
of _____.

    a) 6 and 12   b) 10 and 15

    c) 16 and 18   d) 20 and 30

25. Which organ system is responsible for the production
of sperm?

    a) endocrine system   b) female reproductive system

    c) male reproductive system   d) digestive system

26. Female menstrual cycles normally last about _____.

    a) two days   b) a week

    c) a year   d) a month

27. The human life cycle always begins with ____.

    a) adulthood   b) childhood

    c) puberty   d) adolescence

28. Which one of the following organs produces sperm?

    a) ovaries   b) scrotum

    c) pituitary   d) seminiferous tubules

29. A zygote is _____.

    a) a fertilized egg   b) childhood

    c) an embryo   d) a fetus

30. Which one of the following is a function of the placenta?

    a) providing nutrients to the fetus

    b) protecting the fetus

    c) providing oxygen to the fetus

    d) cushioning the fetus

31. At which of the following months would the fetus's head shift into the down position?

    a) 7 months   b) 9 months

    c) 8 months   d) 6 months

32. The shortest phase of female menstrual cycle is _____.

    a) ovulation   b) oviduct

    c) ovaries   d) uterus

IMMUNE SYSTEM

1. Disease-causing agents are called _____.

2. List three methods of contracting diseases.

3. The two methods by which bacteria can produce illness include _____.

4. The three methods by which infectious diseases are spread include _____.

5. The human body's defense against pathogens is the ____.

6. What are the four lines of defense for the human body?

7. Explain how mucus helps to protect the body from disease.

8. An elevated body temperature is called a(n) _____.

9. The protein that destroys pathogens is called a(n)
_____.

10. What do the letters in AIDS stand for?

11. Explain how vaccines work.

12. Explain how HIV attacks the immune system.

13. Chemical compounds that cause cancer are known as
_____.

14. What are allergies?

15. A bacterial disease that can cause sterility if untreated
is _____.

a) gonorrhea   b) herpes
c) chlamydia   d) both a and c

16. A virus that hides in the body for a long time and
appears suddenly is called _____.

a) herpes   b) chlamydia
c) syphilis   d) gonorrhea

17. What causes diseases?

a) pathogens   b) fungi
c) cigarette smoke   d) all of the above

18. Which one of the following prevents the spread of Lyme disease?

a) washing hands
b) avoiding disease-carrying ticks
c) avoiding kissing
d) covering your mouth

19. Antibiotics fight infections through the process called ____.

a) killing bacteria   b) preventing viruses
c) killing cells   d) killing T cells

20. Asthma is an example of _____.

a) cancer
b) the immune system overreacting to antigens
c) the immune system attacking the body
d) a virus attacking the body

21. Sneezing, runny nose, and itchy eyes due to allergies are caused when the _____.

a) immune system overreacts to antigens
b) cells release histamines
c) antibodies respond
d) pathogens react

22. HIV spreads in the body by _____.

    a) replicating inside the cells of the immune system
    b) stopping antibodies from producing
    c) producing an immune system
    d) all of the above

23. Cancer cells affect the body through _____.

    a) the antibody immune system
    b) taking nutrients needed by other cells
    c) the development of antibodies
    d) the production of pathogens

24. HIV weakens the immune system by killing _____.

    a) B cells   b) antibody cells
    c) helper T cells   d) killer cells

25. HIV is spread by ____.

    a) sexual contact   b) contact with infected blood
    c) sharing of drug needles   d) all of the above

26. What causes cancer?

    a) asthma   b) bacteria
    c) allergies   d) radiation

ANSWER KEY

Life Science

1. Science is a way or process used to investigate what is happening around us. The study of science helps us answer the how, what, where, and why of our surroundings. Science is a way to solve problems.

2. Life science is the study of living things (plants and animals). It helps to explain how living things relate to one another and to the environment.

3. Science and technology influence our lives in many ways. Things such as medicines, vaccines, airplanes, foods, computers, cell phones, and others are the result of science and technology.

4. A scientist is a person who looks for answers to the questions "how, what, where, and why" about our surroundings. Scientists seek to solve problems.

5. Scientists use prior knowledge to make predictions about possible outcomes or answers.

6. Scientists search for clues to help them solve a problem. Scientists may use experiments to help them find the answers to their questions.

7. Scientists use observations, reasoning, classification, and prediction skills to complete their investigations.

8. Science journals are used to communicate scientific observations and experiments to other scientists and the public. They also act as records of a scientist's work.

Scientific Method

1. A variable is any factor tested in an experiment.

2. A manipulated or independent variable is something that is changed by a scientist or intentionally changed within the experiment.

3. A variable that might be affected as a result of that change or the thing that you measure (time, temperature, height, etc.).

4. A variable that is not changed (constant).

5. What you are trying to find out in the experiment, expressed in the form of a question.

6. Your prediction or educated guess.

7. All of the items you use for your experiment are materials.

8. Procedures are how you did the experiment, step by step.

9. What you saw happen, using charts, graphs, pictures, and diagrams.

10. Restate your hypothesis and explain whether the results of your experiment followed your prediction or not. If it did, give examples from your experiment that show your hypothesis predicted the results. If the results did not match your hypothesis, say so and explain what really happened.

11. A process like the scientific method that involves backing up and repeating an experiment.

12. "Scientific method" refers to the body of techniques for investigating phenomena, acquiring new knowledge, or correcting and integrating previous knowledge.

13. Form a hypothesis, test your hypothesis, analyze your data, and draw conclusions.

14. A microscope has one or more lenses that make an enlarged image of an object.

15. For illustration, see Appendix A.

16. A compound light microscope lets light pass through an object and then through two or more lenses. Compound light microscopes also magnify organisms or parts of an organism, making details of structures visible.

17. Stereo microscopes are used to look at thick structures that light can't pass through, such as whole insects or leaves, or your fingertip. It also gives you a three-dimensional view of an object.

18. Things that are too small to be seen with a light microscope can be viewed with an electron microscope. Electron microscopes use a magnetic field to bend beams of electrons. Electron microscopes can magnify images up to one million times.

19. Transmission electron microscopes (TEM) provide images that show great detail.

20. Scanning electron microscope (SEM) transmissions show great detail of the surface of an organism. They are also used to see the surfaces of whole objects.

21. Robert Hooke, an English scientist, made a very thin slice of cork and looked at it under his microscope in 1665.

22. Cells were too small to be seen without the microscope.

23. Improved lenses in compound microscopes allowed individual parts of cells to be seen more clearly. Stereo microscopes allow for three-dimensional viewing. Greater magnification is possible with electron microscopes.

24. For illustration, see Appendix A

Questions about the Scientific Method

1. d) using their feelings about the surroundings

2. d) data

3. d) a single variable

4. c) enable them to be tested

5. d) experiment

6. d) all of the above

Questions about Microscopes

1. c) compound light microscope

2. d) centrifuge

3. b) compound light microscope

4. total magnification

5. magnetic field

6. scanning

7. Zacharias Janssen

8. Theodor Schwann

9. Robert Hooke

10. Rudolf Virchow

Science & Safety Procedures

1. d) if in doubt about any part of an activity, trust your instincts

2. c) wash hands thoroughly after completing the activity

3. d) all of the above

4. a. always obtain your teacher's permission to begin an investigation

b. use the safety equipment provided for you (goggles and safety apron)

c. always hold test tubes away from yourself and others when heating them

d. never eat or drink in the lab and never use lab glassware as food or drink containers

e. any chemical spills should be washed off immediately with water and reported

f. always wash your hands thoroughly after working in the laboratory

g. report any incident or injury

h. always clean your work area before leaving the lab

i. turn off the water and gas, and disconnect electrical devices before leaving the lab

The Science of Biology

1. a) living things

2. a) careful observations

3. b) biologist

4. c) the use of data to support a particular point of view

5. d) gram

6. d) meter

7. a) ounce

8. The metric system is a decimal system of measurement whose units are based on physical standards and scaled in multiples of 10.

9. An inference is a logical interpretation based on prior knowledge or experience. An observation involves using the senses to gather information.

10. With SI, scientists can compare and repeat each other's experiments no matter where they are in the world. They also have a tool for measurement in common.

11. a) liter

The Chemistry of Life

1. b) $C_6H_{12}O_6$

2. c) lipid

3. d) inorganic

4. c) protons and neutrons

5. d) protons, neutrons, and electrons

6. a) sharing electrons

7. b) of water and undissolved material

8. c) evenly distributed mixture of two or more substances

9. b) carbohydrates

10. b) to store and transmit heredity

11. c) chemical reaction

12. d) often happen spontaneously

13. b) catalyst

14. a) water

15. d) solvent

16. b) acid

17. a) electrons

18. b) valence

19. c) positive

20. electrons

21. acid

22. H+ ions

23. making materials that cells need, releasing energy and transferring information, regulating chemical pathways

24. A molecule is the smallest unit of most chemical compounds.

25. A mixture is material composed of two or more elements or compounds that are physically mixed together but not chemically combined.

26. carbohydrates, lipids, nucleic acids, and proteins

27. ionic

28. enzyme

29. d) carbohydrates

30. b) to store and transmit genetic information

Exploring Living Things

1. a) development

2. c) stimulus

3. a) water

4. b) adaptations

5. b) Pasteur

6. c) ability to move in response to the environment

7. b) homeostasis

8. b) living things

9. c) movement

10. carbohydrate

11. made of cells, ability to reproduce, ability to grow and develop, ability to use energy, ability to respond to the environment, ability to maintain a stable internal environment

12. Homeostasis is the process by which organisms keep their internal conditions fairly constant.

13. Metabolism is the combination of reactions through which an organism builds up or breaks down material as it carries out its life processes.

14. cells

15. life span

16. organisms

17. development

18. adaptation

19. Energy is needed for all the life functions, growth, movement, and so on. Without energy, these processes could not take place and organisms would eventually die.

20. A pigeon is made up of cells that use energy to allow it to fly and breathe. It displays movement by flying or hopping. A pigeon jumps at a loud sound or responds to food. It also maintains a constant body temperature, reproduces, grows, and develops.

Cell Structure & Function

1. a) nucleus

2. d) all plants and animals are made of cells

3. d) all of the above

4. a) cytoplasm

5. a) lysosome

6. d) ribosome

7. d) to regulate which materials enter and leave the cell

8. a) work together to perform a specific function

9. b) tissue

10. a) Robert Hooke

11. b) cell membrane

12. b) nucleus

13. b) eukaryotes

14. a) support and protect the cell

15. d) all of the above

16. See illustration in Appendix A.

17. organelles

18. diffusion

19. a. It contains the cell's genetic material in the form of
DNA.

b. It controls many of the cell's activities.

20. Prokaryote cells are generally simpler and smaller than
eukaryotes; eukaryotes have a nucleus and other specialized organelles.

21. cells

22. chromosomes

23. cell membrane

24. It regulates what enters and leaves the cell and provides
protection and support.

25. They are found in plants, algae, fungi, and many prokaryotes.

26. It provides support and protection for the cell.

27. Mitochondria are organelles that convert the chemical
energy stored in food into   compounds that are more
convenient for the cell to use.

28. Chloroplasts are found in plant cells and some other
organisms, but not in animal cells.

29. a. Ribosomes are small particles of RNA and protein
that produce protein following   instructions from the
nucleus.

   b. The endoplasmic reticulum is an internal membrane
system in which components of cell membranes and
some proteins are constructed.
   c. The Golgi apparatus is a stack of membranes in which
enzymes attach carbohydrates and lipids to proteins.
   d. A lysosome is filled with enzymes used to break down
food into particles that can be used by an organism.
   e. A vacuole is a sac-like structure that stores materials.

Cell Processes, Cell Growth & Division

1. d) protons and neutrons

2. b) diffusion

3. c) nucleic acids

4. b) active transport

5. b) glucose

6. a) consumers

7. c) fermentation

8. b) interphase

9. d) only animals have centrioles

10. c) 23

11. a)sex cells from two parents combine

12. c) regeneration

13. b) 23

14. a) haploid

15. b) two

16. c) four

17. c)thymine

18. a) amino acids

19. c) RNA

20. b) prophase

21. b) prophase

22. b) they stop growing

23. a) ATP

24. c) chlorophyll

25. osmosis

26. heterotrophs

27. mitosis

28. ATP is one of the principal compounds that cells use to store and release energy. Energy is released when the chemical bond between the second and third phosphate is broken.

29. carbon dioxide + water —> sugars + oxygen

$$6CO_2 + H_2O \longrightarrow C_6H_{12}O_6 + 6O_2$$

30. oxygen + glucose —> carbon dioxide + water + energy

$$6O_2 + C_6H_{12}O_6 \longrightarrow 6CO_2 + 6H_2O + \text{energy}$$

31. prophase, metaphase, anaphase, telophase

32. centromere

33. The cell cycle is the series of events that cells go through as they grow and divide.

34. See illustration in Appendix A.

35. centriole

36. spindle

37. DNA

38. gene

39. mutation

40. Spindle fibers pull chromosome strands to opposite poles.

41. ATCCGTC is matched by TAGGCAG

42. uracil

43. interphase

44. Cancer is a disorder in which some of the body's own cells lose the ability to control growth.

45. cytokinesis

46. mitosis

47. telophase

48. b) volume increases faster than its surface area

49. a) excess oxygen
50. b) cell division increases the mass of the original cell

51. c) only during metaphase

52. c) in the S phase

53. a) the cell grows

54. c) metaphase

55. a) interphase, prophase, metaphase, anaphase, telophase

56. a) it breaks down the nuclear membrane

57. c) mitosis and cytokinesis

58. c) growth rate

59. Cell division is the process that causes a cell to divide into two new daughter cells.

60. prophase: The chromatin condenses into chromosomes, the centrioles separate (in animal cells) and the nuclear envelope breaks down.

metaphase: The chromosomes line up across the midline of the cell and each chromosome is attached to a spindle fiber and centromere.

anaphase: The sister chromatids separate into individual chromosomes.

telophase: The chromosomes move to the opposite sides of the dividing cell and two new nuclear envelopes form.

61. See illustration in Appendix A.

62. (Answers may vary)

63. When a cell grows larger, more demands are placed on its DNA and the cell has more trouble moving enough nutrients and wastes across the cell membrane.

64. Chromosomes are not visible because the DNA and protein molecules that make up the chromosomes are spread throughout the nucleus.

65. Food molecules are broken down and energy is released in the mitochondria.

66. cell membrane

67. Golgi bodies

68. The gel-like material inside the cell membrane and outside the nucleus is called the cytoplasm.

Heredity & Genetics

1. genetics

2. heredity

3. genotype

4. recessive

5. alleles

6. sex-linked gene

7. They are the sex cells.

8. probability

9. Geneticists use Punnett squares to predict and compare the genetic variations that will   result from a cross.

10. The way in which the alleles segregate is completely random, as in a coin flip.

11.

	T	t
T	TT	Tt
t	Tt	tt

12. Each of the chromosomes in the set from the male parent has a corresponding   chromosome from the female parent.

13. multiple

14. Polygenic inheritance occurs.

15. incomplete dominance

16. Students who are female should answer XX. Male students should answer XY.

17. The chances are zero because the man only passes the Y chromosome to his sons. The gene for color blindness is on the X chromosome.

18. For each son, the chances are 1 in 2, or 50 percent.

19. "Homozygous" means that both alleles for the trait are identical. "Heterozygous" means that the alleles for the trait are different.

20. "Multiple alleles" means having more than two alleles, or types of genes, for a trait. The trait is controlled by one gene pair. Polygenic inheritance involves more than one gene pair controlling the trait. These gene pairs combine to produce the trait.

21. Mr. Mendel described how inherited traits in plants can be transmitted from parents to offspring and he discovered that these traits separated when the plant reproduced.

22. The blood types are A, B, AB, and O. A child with type O blood cannot have one or both parents with type AB blood; he or she can have parents with all other types as long as each has the O allele. A child with type AB blood cannot have one or both parents with type O; he or she cannot have both parents with type B or both parents with type A blood.

Genetic study guide

1. a) genes

2. b) on the X chromosome

3. b) alleles

4. a) traits

5. b) recessive genes

6. a) polygenic inheritance

7. b) an X chromosome

8. a) predict

9. b) the inheritance of traits

10. c) genes

11. a) would have the same phenotypes

12. c) some alleles are dominant and others are recessive

13. c) homozygous

14. b) incomplete dominance

15. d) all organisms

16. a) meiosis

17. b) frequencies of crossing over between genes

18. b) 6

19. a) predict the traits of the offspring produced by genetic crosses

20. b) inherited through the passing of factors from parents to offspring

21. A diploid cell has two sets of chromosomes.

22. The number of chromosomes is reduced by half.

23. a) Mendel

24. c) pedigree

25. b) homologous

26. b) polygenic inheritance

27. a) genetic disorder

28. b) clotting

29. c) producing medicine

30. a) multiple alleles

Evolution

1. Genetics tend to provide the mechanism for evolution to work in a population. If traits were not inherited, there could be no evolution as we understand it.

2. The fittest are those whose adaptations match their environment. Their offspring inherit the adaptive traits and thus are more likely to survive.

3. It would be selected against and die or be killed by a predator because it is too easy to see.

4. species

5. natural selection

6. population

7. Lamarck thought traits acquired by adults were passed to offspring. Darwin thought that only certain traits were inherited or passed on to offspring. (Wallace also thought so.)

8. fossils

9. sedimentary rock

10. past

11. homologous

12. embryology, DNA, homologous structures, vestigial structures, fossils

13. Since DNA is passed from organism to organism, shared ancestors should have similar, but not exactly the same, DNA.

14. Homologous structures, similar DNA, and fossils suggest a common ancestry for primates.

15. Australopithecus, an early example of a hominid, had a small brain case but humanlike jaws and teeth. This is important because it gave scientist information about early man's characteristics that they could use to compare with current human information.

16. They all belong to the order of primates.

17. Today, most extinctions are the result of human interven-
tion. Some people may agree that humans should preserve
species. Others may think that preservation should be left
up to nature.

18. The giraffes with longer necks (a variation) survived by
outliving shorter-necked individuals because they could
eat leaves higher up on trees. In other words, the longer necks were naturally selected for. The variation was
passed on to offspring because it is a genetic trait.

19. The process to figure out the age of a fossil:

    a. note the layer of rock where it is found
    b. do radioactive dating of the fossil
    c. compare the fossil to other fossils
    20. d)   evolution

21. b) radioactive elements

22. c) organisms producing more offspring than can survive

23. a) adaptation

24. b) surviving and reproducing

25. d) endangered species

26. a) Cro-Magnon humans

27. b) natural variation and natural selection

28. b) Thomas Malthus

29. a) James Hutton

30. a) have an innate tendency toward complexity and perfec-
    tion

Classification of Living Organisms

1. Classification is the grouping of ideas, information, or
   objects based on their similarities.

2. Nucleus

3. Plants makes their own food.

4. The four kingdoms of life are: Protists, Fungi, Animals, and
   Plants.

5. The Linnaean System of classification includes classifying
   organisms according to body structures, shape and color,
   and eating system.

6. Taxonomy is a system of grouping and naming living things.

7. The difference between the animal kingdom and the fungi
   kingdom is that animals always move (physically).

8. b) genus

9. a) obtain food from another source

10. a) cells in the environment and our bodies that perform many functions

11. a) genus species

12. a) phylum

13. c) animals and plants

14. b) animalia

15. a) make their own food

16. a) fungi

17. d) large and small categories of living things

18. c) mammalian group

Bacteria & Viruses

1. Bacteria are very small, one-celled monerans.

2. Viruses are particles of nucleic acid and protein that reproduce by infecting living organisms.

3. All viruses possess the ability to enter an organism's cells and use the machinery of infected cells to produce more viruses.

4. When viruses enter the inside of a cell, the viral genes take over and attack the cell.

5. The human body protects itself by producing antibodies and white cells that destroy viruses that attack the body.

6. Pathogens are diseases that produce agents that attack the body.

7. Antibiotics are compounds that block the growth and repro-duction of bacteria.

8. Bacterial cells differ from animal cells in two ways: they do not contain a nucleus or Golgi apparatus.

9. Communicable diseases can be spread through the air, drinking water, touching, and sexual contact.

10. b) develops resistance against antibiotics

11. d) a Golgi apparatus

12. a) fission

13. b) they are eukaryotic

14. b) prokaryotes

15. c) capsule

16. a) flagellum

17. d) all of the above

18. c) tetanus

19. b) it causes the body to produce immunity to fight disease

20. a) strep throat

21. a) DNA or RNA surrounded by a protein cell

22. d) all of the above

23. See illustration in Appendix A.

Plants

1. Plant seeds are dispersed in three ways:

   a. by animals that eat them
   b. through dispersal by the wind
   c. by floating on moving water

2. chlorophyll

3. $C_6H_{12}O_6$

4. An ethnobotanist is someone who specializes in understanding the relationships between people of various cultures and the plants they use.

5. A plant is an organism that is made up of many cells, has chlorophyll, and can make its own food.

6. phloem

7. cambrium

8. stamen

9. pistil

10. stigma

11. stomata

12. During photosynthesis, the plant uses energy from sunlight to convert water and carbon dioxide into high-energy carbohydrates.

13. minerals

14. stem

15. The three basic needs of plants are (a) sunlight, (b) water and minerals, (c) gas exchange.

16. pollination

17. The three structural organs of seed plants are (a) root, (b) stem, (c) leaves.

18. A fruit is a thick wall of tissue that surrounds an angiosperm seed.

19. carbon dioxide + water + light —> sugars + oxygen

20. (a) roots hold plants in the ground, (b) roots absorb water and dissolve nutrients

21. (a) it transports substances between roots and plants, (b) it supports leaves up toward the sunlight

22. We can estimate the age of a tree by counting the rings in a cross section.

23. Transpiration is the loss of water from a plant through its leaves.

24. seed coat

Worms & Mollusks

1. (a) shell, (b) mantle, (c) foot

2. the mantle

3. The shell is made by glands in the mantle that secrete calcium.

4. Land snails and slugs live mainly in moist environments by using a mantle cavity that has a large surface area lined with blood vessels. The lining must be kept moist for oxygen to diffuse across.

5. body walls

6. Most roundworms reproduce bisexually.

7. Hookworm eggs usually hatch outside the body of a host and develop in the soil.

8. scolex

9. attach to the intestinal wall of its host

10. A hermaphrodite worm is one that has male and female reproductive organs.

11. In fission, an organism splits in two and each half grows new parts to become another complete organism.

12. Earthworms improve the soil by fertilizing it with their feces. They also create holes that help plants to absorb water and nutrients.

13. The earthworm has five chambers that act as its heart.

14. gizzard

15. (Answers will vary.)

Fish, Amphibians & Reptiles

1. Fish exchange gases by pulling oxygen-rich water in through their mouths. Then they pump it over their gill filaments where oxygen-poor water is pushed out through slits.

2. The fish bladder adjusts the fish's buoyancy to prevent it from sinking.

3. (a) fins keep fish on course and adjust direction, (b) fins give fish the ability to apply energy for more speed

4. An amphibian is an animal that lives part of its life in water and part of it on land.

5. (a) they change their skin colors for camouflage, (b) they use their skin glands to release unpleasant-tasting skin toxins

6. (a) they have lungs, (b) they have scaly skins, (c) they have eggs that have several protective membranes

7. A reptile is an animal that has dry, scaly skin and can live on land.

8. The reptile's lungs have more gas exchange area than those of amphibians.

9. (a) salamanders, (b) frogs, (c) toads and caecilians

10. Reptile skin does not grow when the rest of the reptile grows. However, reptiles overcome this disadvantage by shedding their skins periodically as they mature.

11. a)gills

12. a) eliminating ammonia from the gills and kidney

13. Amphibians are dependent on water for reproduction and for the development of their fishlike embryos.

14. a) scales

15. b) lungs

16. c) strong bones in the pelvic area

17. b) swim bladder

18. b) amphibians

19. b) reptiles

20. a) lungs

21. c) three chambers

22. a) they do not have shells

23. b) internal fertilization, and they are oviparous

24. c) alligators

25. d) cold climates

26. c) generates its own body heat

27. a) sweating or panting

Ecosystem

1. (a) ponds, (b) forests, (c) meadows, and (d) lakes

2. Palm trees cannot survive in extremely cold temperatures or in areas with little rainfall.

3. A succession is the series of changes a community goes through as it gets older.

4. The organisms that live in an older pond's shallow water are frogs and turtles.

5. An ecosystem is a community interacting with the environment.

6. A greenhouse gas is an atmospheric gas such as CO_2 or methane.

7. Ecological resources include any necessity of life for animals, such as light, food, and water.

8. During the primary succession, pioneer plants help rock to break up for soil formation. They also provide organic material, which assists further plant growth.

9. c) standing water ecosystems

10. c) lava flow

11. a) secondary succession begins on soil and primary succession begins on new surfaces.

12. d) all of the above

13. a) rivers

14. d) temperature ranges

15. b) chemosynthetic bacteria

16. c) autotroph

17. a) omnivores

18. c) third-level consumer

19. b) producers

20. d) population

21. b) sunlight

22. a) ecology

23. b) heterotrophs

24. c) herbivore

25. A food chain is a series of steps in which organisms transfer energy by eating each other, while a food web is a feeding relationship among various organisms in an ecosystem. A food web links together all the food chains in an ecosystem.

26. herbivores

27. See illustration in Appendix A.

28. Commensalism is a relationship in which two organisms live in a community where one benefits while the other gets no benefit but is not harmed. Commensalism occurs between many living things (i.e., orchid plants and trees).

29. c) predator

30. a) habitat

31. primary succession

32. pioneer

33. climax community

The Skeletal System

1. The bones and other connective tissues such as cartilage and ligaments make up the skeletal system.

2. (a) supports the body, (b) protects internal organs, (c) allows movement, (d) stores mineral reserves, (e) provides area for blood cell formation

3. osteoblasts

4. phosphorus and calcium

5. cartilage

6. A joint bone is a place where one bone attaches to another.

7. (a) cardiac, (b) skeletal, (c) smooth

8. Smooth muscles move food through the digestive tract and control the way blood flows.

9. The axial skeleton consists of the skull, vertebral column, and rib cage. The appendicular   skeleton consists of the arms, legs, bones of the pelvis, and shoulder area.

10. Regular exercise increases muscle size and strength. Regular exercise also helps body systems such as organs, heart, and lungs to become more efficient.

11. The five major functions of skin are:

  a. regulating body temperature
  b. excreting wastes, sodium chloride, and water through sweating
  c. serving as a sensory organ for temperature, touch, and pressure
  d. developing vitamin D in the presence of sunlight
  e. forming a protective covering over the body to prevent injury to other body organs

12. c) 206

13. b) minerals

14. d) skeletal

15. d) protecting the brain

16. d) skeleton

17. c) cartilage

18. b) epidermis and dermis

19. d) more than 50 percent

20. c) relaxing the muscles

21. b) hinge

Nutrition & Digestive System

1. Antioxidants are substances found in certain foods that help prevent damage to our bodies.

2. Vegetables and fruits contain vitamins and minerals necessary for energy and growth. They include some nutrients that limit the risk of getting various diseases.

3. Squashes and blueberries are good sources of antioxidants.

4. mechanical

5. The digestive system includes the mouth, pharynx, esophagus, stomach, small intestine, and large intestine.

6. The body loses water through sweat. The water in sweat tends to cool the body as it evaporates from the body. More water is also removed during urination.

7. calories

8. Nutrition is the study of food and its effects on the body.

9. The five nutrients that the body needs are: (a) carbohydrates, (b) water, (c) proteins, (d) fats, (e) vitamins.

10. Food supplies the raw materials used to develop and repair body tissues. It also produces protein that regulates cellular reactions.

11. The function of the digestive system is to convert foods into simpler molecules that can be absorbed and used by the body's cells.

12. Most chemical digestion takes place in the duodenum.

13. The liver produces bile, which helps to dissolve and disperse fat found in fatty foods. Afterward, enzymes break down the food.

14. The primary job of the large intestine is to remove water from undigested material.

15. b) whole grains

16. a) mouth

17. b) vitamins

18. b) red meats

19. a) digestion

20. a) chemical digestion

21. a) six

22. b) proteins

23. a) nutrients

24. a) vitamin A

25. a) grains

Circulatory System

1. (a) heart, (b) blood vessels, (c) blood

2. capillaries

3. (a) avoid smoking, (b) eat a balanced diet, (c) exercise regularly

4. Blood pressure is the force of the blood on the walls of the arteries.

5. veins

6. pulmonary circulation

7. b) brain

8. c) lungs and heart

9. a) hypertension

10. c) bone marrow

11. b) circulatory system

12. a) capillaries

13. b) red blood cells

14. d) air passage

15. b) systemic circulation

16. a) they prevent the backward flow of blood

17. The circulatory system is the set of vessels and muscles that help control the flow of blood around the body. The main parts of the system are the heart, arteries, capillaries, and veins.

18. (Answers may vary.)

Respiratory & Excretory System

1. The functions of the urinary system are:

    a. to rid the body of waste
    b. to help control blood and remove excess water
    c. to remove excess salt from the body

2. (a) skin, (b) lungs, (c) kidneys, (d) liver

3. excretion

4. The excretory system is the body system made up of those organs that rid the body of liquid waste.

5. dialysis

6. kidney

7. urinary bladder

8. The ureter carries urine from the kidneys to the bladder.

9. The body loses water through sweat and urine.

10. homeostasis

11. The respiratory system brings about the exchange of oxygen and carbon dioxide.

12. medulla oblongata

13. Passengers in a depressurized airplane cabin must breathe more oxygen because they have more CO_2 in their blood than usual.

14. breathing

15. alveoli

16. Respiration is the process of oxygen and carbon dioxide exchange between the lungs and the environment.

17. A small piece of cartilage called the epiglottis covers the entrance to the trachea to prevent food from entering it when you eat.

18. When you inhale, the diaphragm contracts and the rib cage rises. This process expands the volume of the chest and creates a vacuum and atmospheric pressure that causes the lungs to fill with air.

19. bronchioles

20. Carbon monoxide is dangerous because it blocks the transport of oxygen by hemoglobin, preventing the heart and the other vital organs from getting the oxygen they need to function.

21. Emphysema is the loss of elasticity in the lungs.

22. (a) emphysema, (b) lung cancer, (c) chronic bronchitis

23. passive smoking

24. It is difficult to quit smoking because nicotine is a power-ful and addictive drug.

25. true

26. The functions of the following are:

a.trachea: brings air from the nasal chamber to the bronchi
b. bronchi: carry air from the trachea to the lungs
c. alveoli: provide gas exchange between the lungs and blood

27. As you breathe in, your rib cage moves up and out as the diaphragm moves down.

28. When someone eats, food may fall into the open windpipe and cause him or her to choke.

29. c) carbon dioxide

30. b) oxygen

31. c) epiglottis

32. d) pharynx

33. b) alveoli

34. b) tar

35. d) all of the above

36. b) nose

37. a) diaphragm

38. d) brain

39. c) coffee

40. d) pneumonia

41. c) skin

42. c) excretory

43. b) circulation

44. b) large intestine

45. a) removing waste

46. c) salt

47. d) pharynx

48. b) excess salt

49. d) all of the above

50. a) dialysis

Nervous System

1. The nervous system controls and coordinates body functions and responds to internal and external changes.

2. Neurons are classified according to the direction an impulse travels within the body.

3. (a) motor neurons, (b) sensory neurons, (c) interneurons

4. (a) brain, (b) spinal cord

5. The central nervous system relays information, processes messages, and analyzes information.

6. A nerve impulse begins when a neuron is stimulated by another neuron.

7. A reflex allows the body to respond to danger immediately without wasting time.

8. A reflex is a process for responding to danger.

9. The nervous system is the body system made of cells and organs that allow organisms to detect changes and respond to them.

10. Sensory receptors are neurons that react directly to stimuli from the environment.

11. (a) pain receptors   (b) thermoreceptors,   (c) chemoreceptors, (d) mechanoreceptors

12. The autonomic nervous system regulates activities that are involuntary or automatic, such as the heartbeat.

13. Drugs disrupt the function of the nervous system by interfering with the action of neurotransmitters.

14. Alcohol slows down the central nervous system.

15. See illustration in Appendix A.

16. Dendrites carry the impulse from the environment or from other neurons toward the body cell.

17. a) synapse

18. c) reflex

19. a) dendrite

20. a) sensory neurons

21. d) motor neurons

22. c) interneurons

23. c) cerebrum

24. c) spinal cord

25. b) homeostasis

26. d) all of the above

27. b) stimulants

28. c) tissue

29. d) peripheral

30. b) to control activities of the body

31. c) peripheral nervous system

Human Reproduction

1. fertilization

2. vagina

3. zygote

4. placenta

5. fetus

6. ovaries

7. scrotum

8. The testes are located outside the body to maintain a cooler temperature, which is necessary for sperm production.

9. sperm and nourishing fluid

10. once a month

11. Ovulation is the process in which an egg is released from the ovary.

12. Puberty is the time when a person becomes physically able to reproduce.

13. When a woman reaches menopause, ovulation and menstruation stop. The period of time before this occurs is often referred to as perimenopause when ovulation and menstruation begin to gradually cease.

14. Human reproduction is very important because it continues the species.

15. There are 23 chromosomes in each sperm cell.

16. A zygote will inherit 46 chromosomes from its parents.

17. When two eggs are released and fertilized by two different sperm, fraternal twins are the result.

18. b) ovaries

19. b) oviduct

20. c) pituitary gland

21. c) 200–300 million

22. d) 23 chromosomes

23. c) a zygote

24. b) 10 and 15

25. c) male reproductive system

26. d) a month

27. b) childhood

28. d) seminiferous tubules

29. a) a fertilized egg

30. a) providing nutrients to the fetus

31. b) 9 months

32. a) ovulation

Immune System

1. pathogens

2. Disease can be inherited, caused by materials in the environment, and produced by agents.

3. releasing toxins and breaking down tissue

4. (a) coughing and sneezing, (b) contaminated water and food, (c) physical contact

5. immune system

6. (a) sweat, (b) tears, (c) the skin, (d) mucus

7. Mucus helps protect the human body from disease by trapping pathogens.

8. fever

9. antibody

10. Acquired Immune Deficiency Syndrome

11. Vaccines work by causing the body to form its own antibodies.

12. HIV attacks the human immune system by attacking the lymphocytes that normally fight antigens and chemicals. As a result, the body cannot fight off invading antigens.

13. carcinogens

14. Allergies are immune system overreactions to certain antigens. They usually cause sneezing, runny nose, and watery eyes.

15. d) both (a) and (c)

16. a) herpes

17. d) all of the above

18. b) avoiding disease-carrying ticks

19. a) killing bacteria

20. b) immune system overreacting to antigens

21. b) cells release histamines

22. a) replicating inside the cells of the immune system

23. b) taking nutrients needed by other cells

24. c) helper T cells

25. d) all of the above

26. d) radiation

Appendix A

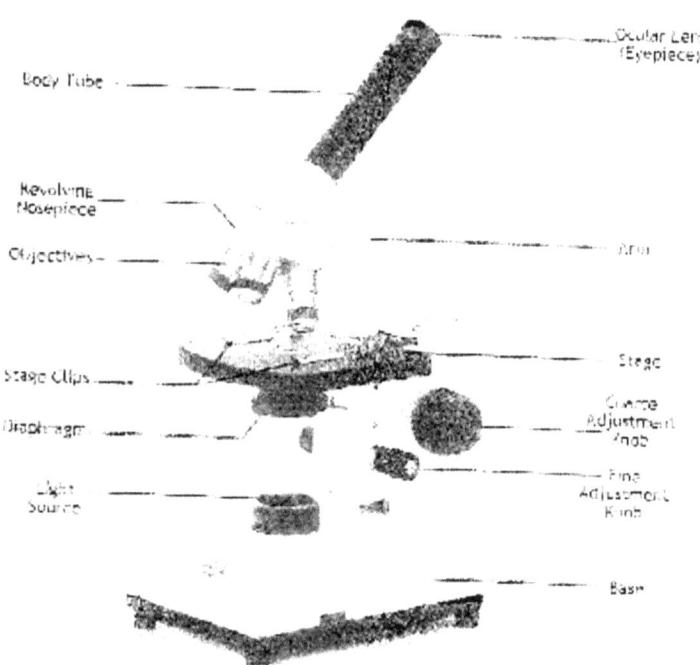

Ocular Lens
(Eyepiece)

Body Tube

Revolving
Nosepiece

Objectives

Arm

Stage

Stage Clips

Coarse
Adjustment
Knob

Diaphragm

Fine
Adjustment
Knob

Light
Source

Base

THE MICROSCOPE

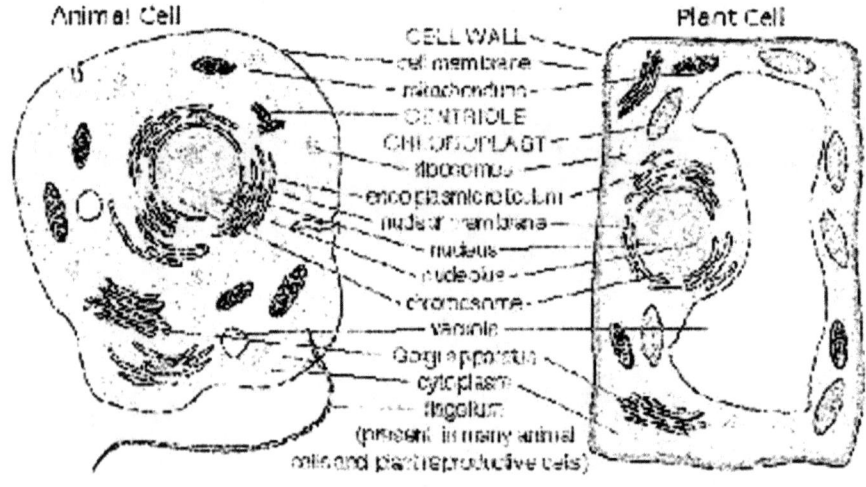

Animal Cell

Plant Cell

CELL WALL
cell membrane
mitochondria
CENTRIOLE
CHLOROPLAST
ribosomes
endoplasmic reticulum
nuclear membrane
nucleus
nucleolus
chromosome
vacuole
Golgi apparatus
cytoplasm
flagellum
(present in many animal
cells and plant reproductive cells)

The 4 Stages of Mitosis

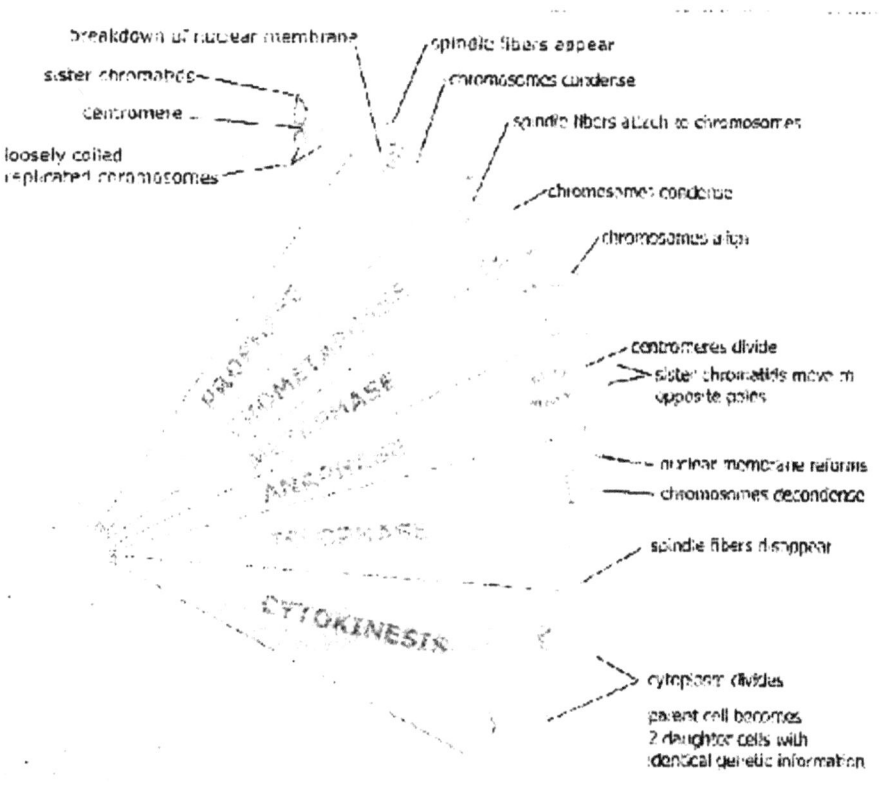

breakdown of nuclear membrane

spindle fibers appear

sister chromatids

chromosomes condense

centromere

spindle fibers attach to chromosomes

loosely coiled
replicated chromosomes

chromosomes condense

chromosomes align

centromeres divide

sister chromatids move in
opposite poles

nuclear membrane reforms

chromosomes decondense

spindle fibers disappear

cytoplasm divides

parent cell becomes
2 daughter cells with
identical genetic information

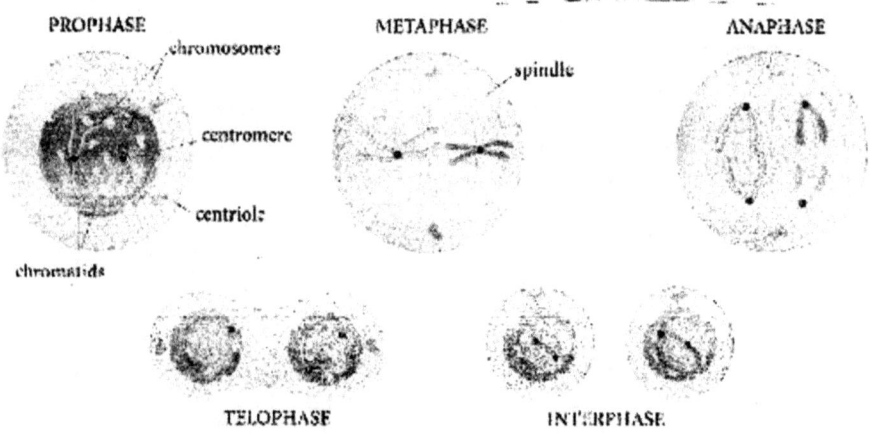

The 4 Stages of Mitosis

PROPHASE — chromosomes, centromere, centriole, chromatids

METAPHASE — spindle

ANAPHASE

TELOPHASE

INTERPHASE

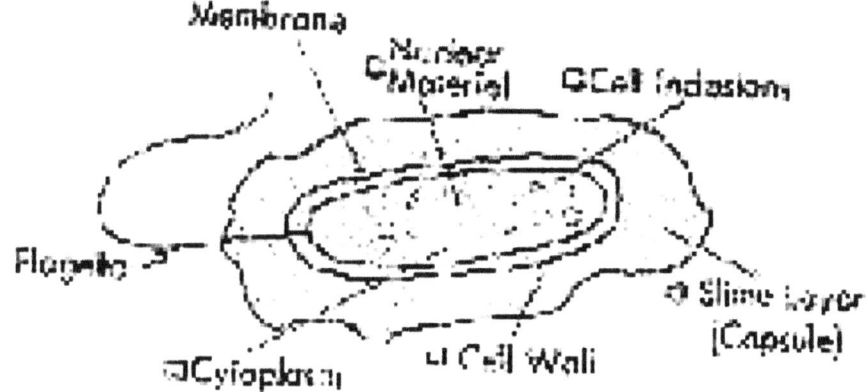

Bacterial Cell

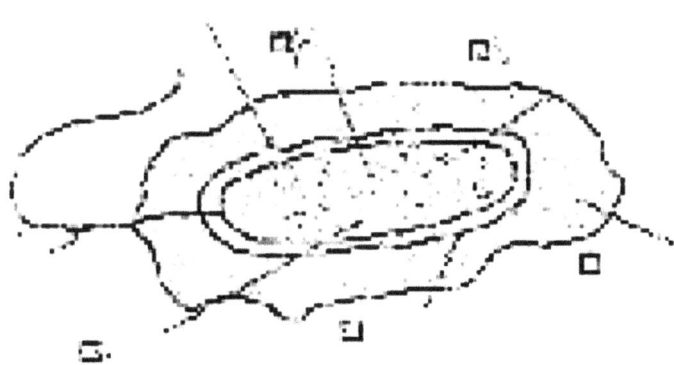

Bacterial Cell

Earthworm

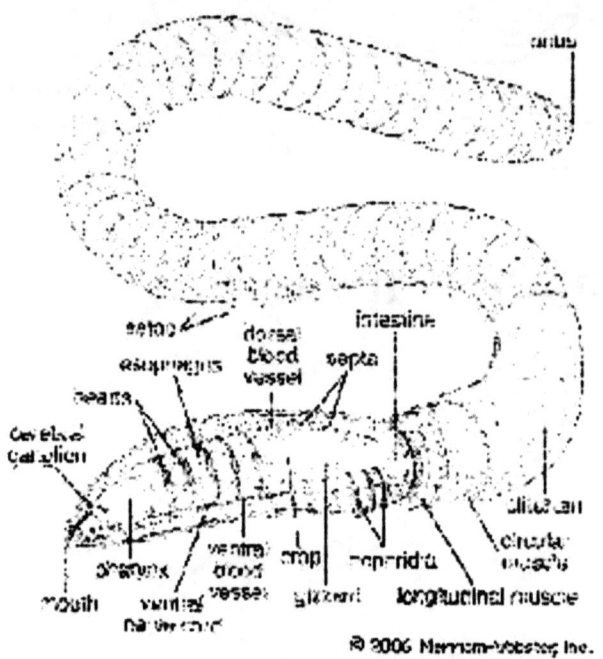

The earthworm

Earthworm Anatomy

Label the parts of the earthworm

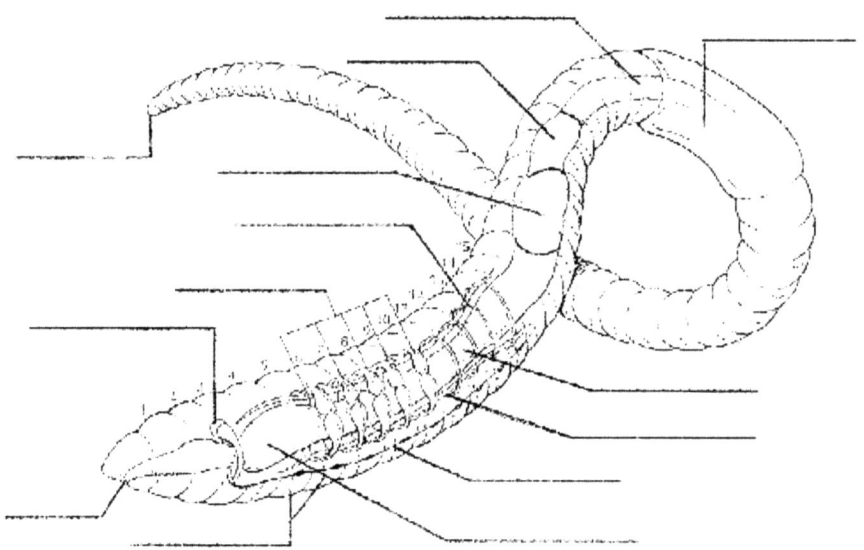

Food Web

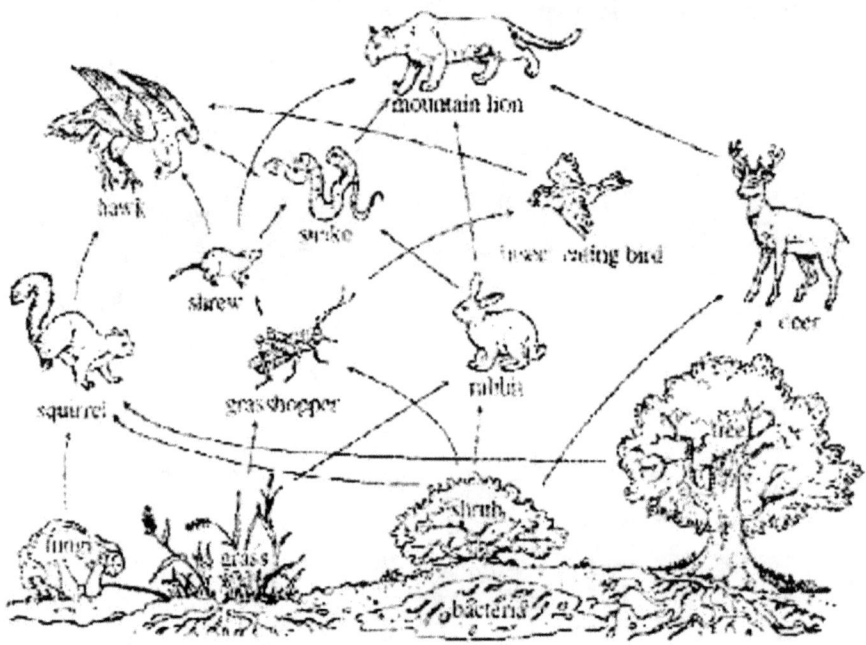

Circulatory System

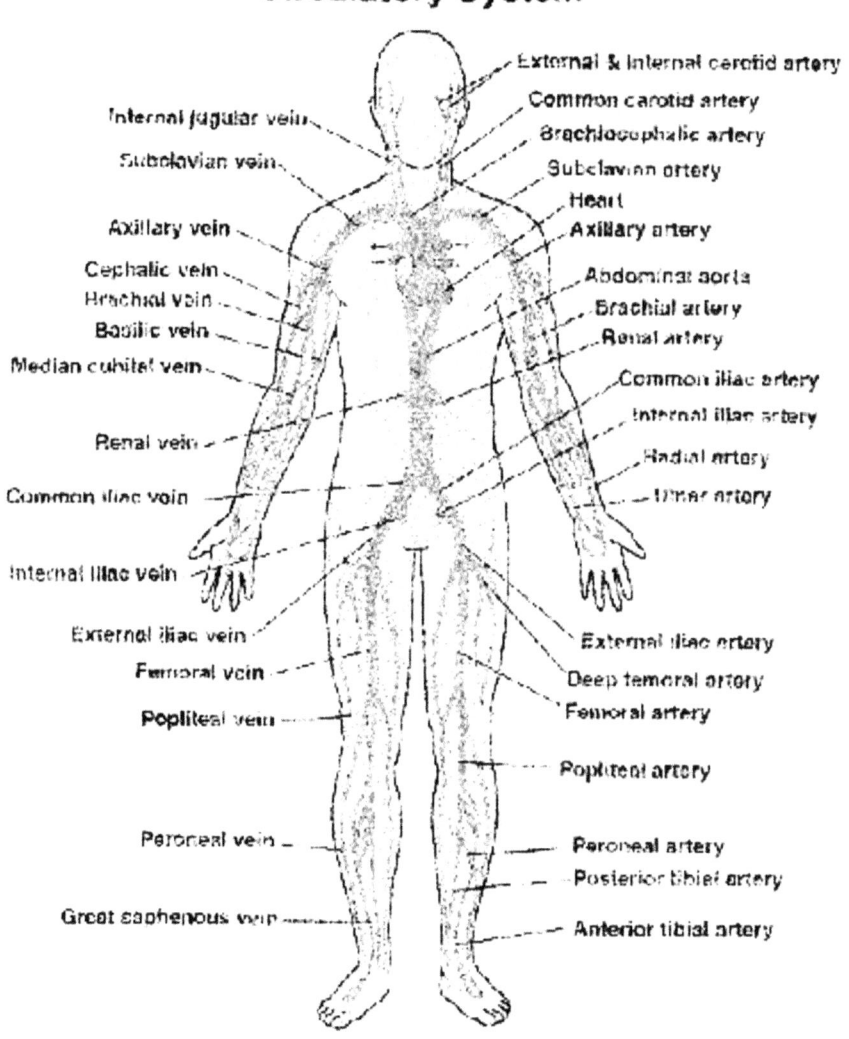

External & Internal carotid artery

Internal jugular vein

Common carotid artery

Brachiocephalic artery

Subclavian vein

Subclavian artery

Heart

Axillary vein

Axillary artery

Cephalic vein

Abdominal aorta

Brachial vein

Brachial artery

Basilic vein

Renal artery

Median cubital vein

Common iliac artery

Internal iliac artery

Renal vein

Radial artery

Common iliac vein

Ulnar artery

Internal iliac vein

External iliac vein

External iliac artery

Femoral vein

Deep femoral artery

Popliteal vein

Femoral artery

Popliteal artery

Peroneal vein

Peroneal artery

Posterior tibial artery

Great saphenous vein

Anterior tibial artery

The Nervous System

The Nerve Cell

The basic unit of the nervous system is the nerve cell, or **neuron**. The basic function of the neuron is to transmit information. There are approximately 28 billion neurons in the human body penetrating every tissue in every part. Neurons vary greatly in size and shape, with the longest ones— those that extend down the leg as part of the sciatic nerve—measuring over one meter. All nerve cells have a similar structure.

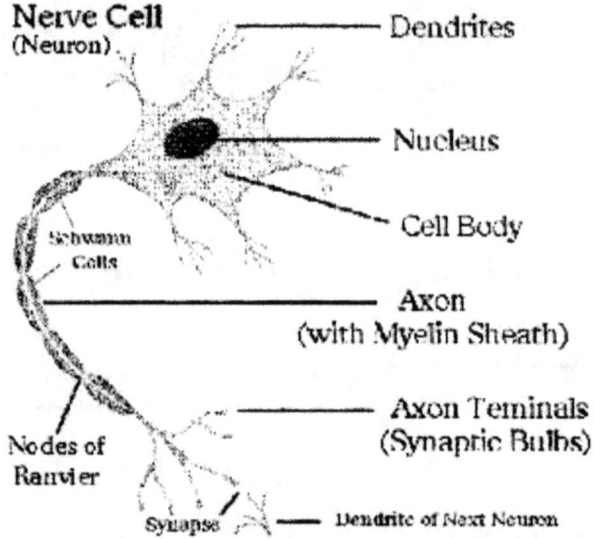

Nerve Cell (Neuron) — Dendrites — Nucleus — Cell Body — Axon (with Myelin Sheath) — Axon Terminals (Synaptic Bulbs) — Schwann Cells — Nodes of Ranvier — Synapse — Dendrite of Next Neuron

www.ingramcontent.com/pod-product-compliance
Lightning Source LLC
Chambersburg PA
CBHW051502170526
45166CB00001B/348